Understanding Microprocessors

Macmillan Computer Science Series

Consulting Editor
Professor F. H. Sumner, University of Manchester

S. T. Allworth, *Introduction to Real-time Software Design*

Ian O. Angell, *A Practical Introduction to Computer Graphics*

G. M. Birtwistle, *Discrete Event Modelling on Simula*

T. B. Boffey, *Graph Theory in Operations Research*

Richard Bornat, *Understanding and Writing Compilers*

J. K. Buckle, *The ICL 2900 Series*

Derek Coleman, *A Structured Programming Approach to Data**

Andrew J. T. Colin, *Fundamentals of Computer Science*

Andrew J. T. Colin, *Programming and Problem-solving in Algol 68**

S. M. Deen, *Fundamentals of Data Base Systems**

J. B. Gosling, *Design of Arithmetic Units for Digital Computers*

David Hopkin and Barbara Moss, *Automata**

Roger Hutty, *Fortran for Students*

H. Kopetz, *Software Reliability*

A. Learner and A. J. Powell, *An Introduction to Algol 68 through Problems**

A. M. Lister, *Fundamentals of Operating Systems, second edition**

G. P. McKeown and V. J. Rayward-Smith, *Mathematics for Computing*

Brian Meek, *Fortran, PL/I and the Algols*

Derrick Morris and Roland N. Ibbett, *The MU5 Computer System*

John Race, *Case Studies in Systems Analysis*

B. S. Walker, *Understanding Microprocessors*

I. R. Wilson and A. M. Addyman, *A Practical Introduction to Pascal*

* The titles marked with an asterisk were prepared during the Consulting Editorship of Professor J. S. Rohl, University of Western Australia.

Understanding Microprocessors

B. S. Walker
Department of Cybernetics,
University of Reading

© B. S. Walker 1982

All rights reserved. No part of this publication may be reproduced or transmitted, in any form or by any means, without permission.

First published 1982 by
THE MACMILLAN PRESS LTD
London and Basingstoke
Companies and representatives
throughout the world

Typeset by Reproduction Drawings Ltd., Sutton, Surrey.

Printed in Great Britain
by Unwin Brothers Limited
The Gresham Press, Old Woking, Surrey

ISBN 0 333 32309 2

The paperback edition of the book is sold subject to the condition that it shall not, by way of trade or otherwise, be lent, resold, hired out, or otherwise circulated without the publisher's prior consent in any form of binding or cover other than that in which it is published and without a similar condition including this condition being imposed on the subsequent purchaser.

Contents

Preface   vii

1. Introduction to Data Processors   1

the concept of a data processor—data processors—the microprocessor—programs and data—the logic—microprocessor configurations

2. A Simple Central Processing Unit   10

arrangement and operation—simple functions—jumps or branches—highways and microprogramming—the arithmetic and logic unit and its function—number representation

3. The Memory or Store   23

types of store—characteristics of microcircuit stores, RAM, ROM, PROM, EPROM, EAROM—addressing strategy and addressing mechanisms—immediate, direct, relative addressing—indirection, modification, indexing, stacks

4. Practical CPUs   38

the amplified CPU—the special registers: pointers, central registers, auxiliary registers, the control and flag registers—comparison of some recent and current microprocessors: Intel 8008 and 8080, Motorola M6800, National SC/MP, Zilog Z80

5. Programming   54

machine code—assemblers: subroutines and macros—high level languages—development of microprocessor programs—emulators, simulators and other aids

6. Interfacing to Peripherals   66

the nature of the problem of microprocessor interfacing—interfacing devices—the parallel interface unit—handshakes, interrupts—serial interface units, UAR/T

7. Microprocessors in Perspective 74

8-bit processors—the single-chip microcomputer—12 and 16-bit microprocessors—bit-slice devices—what of the future?

Appendix A Binary Arithmetic 81
number representation—arithmetic operations—signed numbers

Appendix B Computer Logic 88
introduction—logic elements—combinational logic—sequential elements, flip-flops, etc.—ALU logic—busbar logic devices

Appendix C Transistors and Microcircuits 102
historical development—semiconductors—the MOS transistor

Index 109

Preface

Electronic digital computers and electronic data processors are two names for the same devices. The latter is a better name since it more nearly describes what they do. Microprocessors are physically very small electronic data processors. Although they have similar capabilities to their much larger brethren they are fabricated on a single chip of silicon or sapphire. They must be made in huge quantities to be economical and since they are so made they are cheap. A few years ago it cost a lot of money or academic attainment to join the quite select club of computer users, whereas today anybody can own one or even several. Before long many people, who will have no idea that they are computer users, will own several. They are becoming integral parts in car electrical systems, in domestic automation such as washing machines and they already form the basis of the proliferating numbers of hand-held and desk calculators, electronic TV games and the like.

It is the purpose of this book to try to explain what microprocessors do and how they do it, in so simple a way that people who do not regard themselves as technically inclined can still understand them and develop informed ideas about what the microprocessor may well be able to do and also what it may not do.

For those who are technically minded it is hoped that this book will point the way for them to develop their skills in the highly significant area of automatic systems under microprocessor control.

The principles by which data processors or computers work are elegantly simple—like the wheel—when once understood. It is difficult to imagine why such a simple concept took so long to become obvious. What makes them seem mysterious or difficult to comprehend is that they do things—very simple things—at speeds that we, as humans, find to be incredible because our concepts of speed and time are related to bicycles, cars or aeroplanes but not electrons.

Computers seem to do very complicated things because they do so many simple things so quickly. Yet they are in many respects sluggish compared with the human nervous system, which we take for granted. Even the most advanced man-made control systems are crude by comparison with those in nature. What seem to us quite simple things (because we do not look into them in great detail) like catching a ball, or chewing and swallowing food, or writing a letter, when examined in precise detail do become much more complicated. Processors are artefacts and they need to be dealt with in great detail in many applications; for this reason they too seem complicated. Their designers and builders need to go into this detail, but their users need not.

In order to develop an understanding of automatic computers and data processors the easy approach is to examine them and what they do first in a fairly cursory way and then to examine them in increasingly fine detail—but only as far as we wish, or need, or as our curiosity impels us. This will be the approach in this book. Readers are invited to read just so far as they may feel impelled. Having read all of it, they may continue further, as far as they can possibly go, to the limits of human expertise, if they are sufficiently interested and determined.

The understanding of computer or microprocessor principles does not primarily depend on detailed knowledge of electronics, logic design, binary arithmetic or the processes by which microcircuits are made. This knowledge becomes increasingly necessary, however, in the design and implementation stage of practical systems. For this reason the main text has been written with the minimum stress on these topics. The three appendixes following, however, do go into more detail for the reader who feels the need.

Finally there is a problem of terminology—sometimes called jargon. Like every other technology, that of computers and microprocessors does contain words and expressions which imply specific concepts and which it would be unrealistic and tedious to avoid. Computer technologists generally have taken pains to use ordinary words with their normally understood meanings. Whenever computer terminology is introduced in the text, an explanation is given.

1 Introduction to Data Processors

The concept of a processor is a familiar one. It is a contrivance into which we put some material and it operates on what we put in, producing at its output some modified, hopefully improved, form of the input material. For instance a meat canning factory is a kind of processor, as is a mill for making paper or grinding corn into flour. More than a century ago, Charles Babbage built a 'calculating engine'; he called the 'number crunching' part of his engine the 'mill'.

A data processor fits happily into this same concept of a processor. *Data* is a Latin word and means 'things given'. The term, as used today, means 'abstract things', 'items of information expressed in symbols'. We feed data into an electronic data processor and it operates on the given things, or their symbols; it outputs other modified data in a more useful form.

A typical example of a data processing operation is the preparation of the payroll for the work-force of a factory. The symbols representing the names of the employees are entered as data, together with other data relating to their hourly rates and hours worked, all as symbols. The processor operates on this data and in due course prints out the pay slips for the employees describing, in symbols, their pay, tax deductions, stoppages, bonuses, etc.

A different kind of process might be to read in data describing the behaviour of a reaction in atomic physics and to put out data describing the result, perhaps in symbols or perhaps in the form of a graph. Another different process might be used in a library to keep track of book loans or to find references. Different again, a processor may be used to fly an aeroplane. Here the data is not translated in the first place into discrete symbols but is supplied in some more continuous and convenient form. The processor compares data relating to present speed, course, altitude, attitude, etc., with other data provided by the pilot and sends out data to the aircraft controls which brings the actual values into coincidence with the desired values.

The key to the versatility of the data processor is that it contains a variety of data processing elements. Some of the data which goes in is called the program: this data is itself used to control the activities of the elements, how and when they are used and in what order. The data to be processed is then read in and operated on within the machine; the output is due to both.

The program has to specify in fine detail what the processor does to the processed data. Writing about Babbage's engine, Countess Ada Lovelace (Byron's daughter) remarked that 'the machine can do only that which we know how to order it to do'; what she wrote is equally applicable to the most modern data processors.

Figure 1.1 illustrates crudely the basic organisation of a data processor.

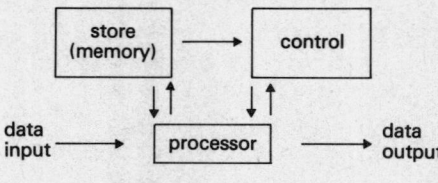

Figure 1.1

Suppose we consider the data processing problem of sorting a mixed-up list of numbered files into a serially numbered list: such a task might be the production of a telephone directory to be listed in number order from an alphabetically ordered one. The processor illustrated in figure 1.1 could perform the job in the following way.

First of all a program would have to be prepared and entered (read in) to the processor. It would necessarily be conceived and written by a human. It would consist of a list of rather simple instructions for the processor to obey in a clearly defined sequence.

Expressing the operation of the program very simply: it would call in the first items of the data to be sorted and store them also in memory. It would then compare the numbers in the first two files, storing the larger as the start of a new list, and keeping the smaller. It would then take in the next file and compare the number in this one with the one it was holding. The larger of these two it would place on the new list, keeping the smaller for the next comparison, and so on. When it had scanned all the original list of data it would repeat the process but on the new list, forming another. It would keep on repeating this process until a scan of the whole list entailed no swopping.

There are many ways in which a 'sort' of this kind can be done by a processor; the one described is neither very elegant nor efficient, but would have advantages in certain circumstances. It does, however, illustrate some important points.

First, it illustrates how the use of a very simple procedure repeated many times can perform a complicated process. If it is done by electronic means at great speed, then even such an inelegant repetitive process can be performed quite quickly measured against the human time scale.

Secondly, it illustrates an important aspect of programming such processes. Quite a simple routine, or list of instructions, would be required to perform the comparison and list. This same short routine would be written once but used over and over again. The same would be the case for the routines for reading in each item, for switching from list to list, for testing the number of comparisons and, finally, for printing out the result.

Thirdly, it illustrates the functions and requirements of various parts of the system. The 'input' would be required to handle the volume of input data at a sufficient speed to supply the processor. The 'memory store' would need to hold the program and some or all of the listed data for some or all of the time. The program, once entered would remain unchanged throughout the process. The data would be continually changing. The 'control' would have to be able to decode and give effect to the program instructions.

Finally the output device would need to print out the symbols which formed the resulting list.

Obviously, careful preconsideration of the problem could make the process more efficient. For instance, the numbers on which the comparisons are performed may well be only a small fragment of each file. It could well be possible for the majority of each file to bypass the processor itself or perhaps not enter the system at all.

THE MICROPROCESSOR SYSTEM

For many people the term 'microprocessor' has come to mean not just the device itself but the whole assembly of things that go with it to form a microprocessor-based system. Hence, to these people, the idea of the microelectronic magic chip is somewhat ahead of what the technology has yet been able to achieve. Most microcomputers and microprocessor-based systems consist of a considerable number of chips and an even greater number and volume of other devices used to maintain, control and harness them.

The term microprocessor, accurately used, means only the *central processing unit* (CPU) of the whole system and contains the following.

(1) The *arithmetic and logic unit* (ALU) which does the 'number crunching', the logical operations, comparisons and the like, on data fed into it.
(2) A group of highly specialised memory storage elements which are used in the ALU activities. These are commonly termed the *central registers*. The term register has the same conceptual meaning in the computer context as in the world outside it: it is a thing which holds (and sometimes displays) a number, or piece of symbolised data.

(3) A *control unit* which has two main functions. The first is to fetch instructions from memory and decode them in the proper order. The second is to execute them, that is to control the ALU and its attendant registers and, when required, fetch its data (*operands* is the technical term). It also has to do a host of less obvious things which are often referred to as 'house-keeping', like refreshing the contents of storage elements, counting operations internally and sequencing what are called 'microinstructions'. The 'simple' operations of the ALU are broken down into even more elementary ones by the designer and formed as 'microprograms' of strings of 'microinstructions'.

The whole CPU assembly is fabricated on a chip of silicon with an area of a few square millimetres. The chip is then stuck to a substrate of ceramic or similar material and connected to a set of terminals by gold wires that are very thin indeed. The terminals are rigidly held to the substrate and the whole assembly is encapsulated to make it sufficiently robust for humans to handle with reasonable care. It does not require much imagination nor technical knowledge to appreciate that 'pin-out' is a limiting constraint on microprocessors. Pin-out is the arrangement of terminals on the encapsulated package, joined by the gold wires to the chip. It is extremely difficult to make microcircuit capsules with 40 terminals, though some do exist with as many as 64. Since the CPU has a great number of things to do and has to receive and output all the data that is processed, much ingenuity is required to make everything possible through the limited number of connections. This is one of the principal differences of microprocessors from larger data processors in which the provision of a few or even a lot of extra connections poses no insuperable problems.

The processor of figure 1.1 points to some of the other things needed for a microprocessor system. The CPU contains the 'control' and 'processor' boxes. The next main item that is needed is the memory or store. The main memory of a microprocessor system is also formed of chips much like the CPU but having very many more elements of fewer varieties. The amount of microcircuit store attached to the CPU depends on the tasks the system must perform, on the pocket of the purchaser and on the pin-out. As we shall see later in more detail, the maximum practicable store size for direct access by the CPU is about 64 thousand (64 k) words or *bytes*, that is, 8-bit patterns or store locations. A memory store is organised like apartments in a housing block. Each location capable of holding a unit of data has an 'address'. To put data in or get it out first we must specify to the store controller the address and then send the data into, or read it out of, that address. There are a variety of different types of microcircuit store and these form a topic for consideration later in this book. But generally even for small systems there will need to be, at this stage of the technology, several chips, or capsules, of memory store.

Figure 1.1 vaguely indicates 'input' and 'output': these labels cover a multitude of devices and the chips to control them. Generally they are referred to as *peripherals* or *peripheral devices* since conceptually they are attached to the periphery of the system. Among other things, they include paper and magnetic tape data transmitters and receivers, teletypes and line printers, keyboards, cathode ray tube monitors for TV-type presentation, data converters of various kinds, lamps, buttons, switches, 'floppy discs', magnetic and punched card readers—and so the list goes on. More generally these are the contrivances which translate humanly comprehensible data into machine data and vice versa, thus providing the 'man-machine interface'.

Finally, not shown, but important nevertheless, there need to be power supplies, cooling fans and similar unromantic provisions. The microprocessor chip when harnessed is usually only a tiny component in a cabinet or rack of electronic and electromechanical devices.

The system cannot operate without a program. This is a list of instructions to the processor detailing every processing action and the order in which these operations are to be performed. In a dedicated system the program may be held in a *read only memory* (ROM), in which case the system should be able to operate with a minimum of external control, perhaps merely a switch or button to set it to run or stop. In order to prepare such a ROM there will have to be a development stage, however, in which the system under development has to be run and tested and the program debugged (that is, corrected where necessary). It is very unlikely that even the simplest program will be written correct in every detail in the first place. Often, in 'real-time' system design, the exact timing and similar details cannot be correctly forecast before the system is run and adjustments must be made experimentally. In these cases the program must be held in a read/write memory until all its details are completely tested and corrected.

In microprocessor systems the read/write memory is usually also made up of microcircuits; it is called *random access memory* (RAM). This somewhat misleading name merely means that the data held anywhere within the memory is equally accessible, that is, it does not have to be read in a serial sequence; accessing data in a random order entails no time penalty.

Program instructions may be of two kinds, those which require operands to be obtained from memory and those which do not. Those requiring data are called *memory reference instructions* and must contain two parts. The first part defines the function to be performed and the second defines the location in the memory store from which the operand or data is to be obtained. Instructions which are not 'memory reference' are sometimes called *operation instructions*: they must also define a function—this may be to alter the state of the processor or some component of it or to manipulate data that is already held within the processor.

Data on which the processor is to operate is normally held in the RAM. Since the processor operates very quickly, if wanted data is not in RAM, the processor will have to wait for it to be read in through some peripheral device or from a bulk store such as a *floppy disc*, a device which records and reads back data in serial format, storing it on a magnetic film on the surface of a plastic disc. The very rapid growth in the use of microprocessors has led to an equally rapid development of this kind of storage device with the result that the floppy disc has become particularly associated with microprocessor systems.

When the data to be processed is too voluminous to be held economically in RAM, much of the success of the system may depend on the efficient organisation of transfers of blocks of data between the bulk store and RAM, enabling the processor to have available the data it needs without having to wait for it. In larger computer systems, RAM is often called *immediate access storage* and the RAM store the *working store*. Discs and similar bulk storage devices are often called *mass store*, while magnetic tape stores and interchangeable disc units are categorised as *backing store*. Since microprocessors, computers and data processors are so similar in function, terminology applicable to one often finds its way into the literature of the others, often very usefully, but it can be confusing. Microprocessors themselves have already led to the development of a special vocabulary; a number of words and expressions used, however, imply important concepts and we must become familiar with them.

The material operated on by microprocessors is data in the form of symbols. The machine has to be able to interpret and act according to instructions, expressed also as symbols. From the engineering considerations alone, it is necessary that the symbology is the simplest. It is for this reason that computer designers from the earliest days have used a binary notation, that is, restricted to the use of only two symbols which the human writes and interprets as 1, or 0, but which may be interpreted also as logically 'true' or 'false', or electrically as 'on' or 'off'.

From an electronic point of view it is far easier to construct devices which indicate two rather than a multiplicity of states. We do not need to understand much about electronics to understand microprocessors. For this reason all but the necessary minimum amount of electronics has, in this book, been relegated to an appendix, as has the detail of logic elements and other devices and techniques which are used in the engineering implementation of the microprocessor. These affect its performance and physical construction but have little to do with the fundamental principles by which it works.

Bits

The symbol space in which we can record a 0 or 1 is called a *bit*. The *bit* is, to the student of information theory, the minimum quantity of information required to resolve the ambiguity of equiprobable alternatives. The computer technologist uses it more loosely as the abbreviated form of the term 'binary digit'. The microprocessors with which this book is primarily concerned operate on arrays of 8 bits, called 8-bit binary words or more shortly 8-bit *bytes* or, by common usage, just 'bytes'. For example

$$0\ 1\ 0\ 0\ 1\ 1\ 1\ 0$$

is a byte of 8 binary digits or bits and is a pattern of 1s and 0s. Such a pattern can be translated or 'transduced' for the human from some electronic representation into, for instance, a row of lamps, lit denoting '1' and unlit denoting '0'. A similar pattern can be formed by punching holes across a strip of paper tape which could be punched by a teletype and transduced by a photoelectric device into electronic signals which the processor can manipulate or interpret.

Registers

An array of electronic devices can be constructed each of which has the capability of being set to 1 or reset to 0, by an electronic signal, and it remains for the time being in that state. Such an array is called a *register* because it registers a pattern, sometimes having the significance of a number. Registers are a fundamental component of the microprocessor.

Logic Elements

The logic values, 1 or 0, can be transferred from one register to another as electrical signals or voltage levels. These transfers are regulated by logic elements or *gates*. All the fundamental operations of the microprocessor are achieved by these transfers. Data is manipulated or changed by the logic units between registers in a logically defined way. For instance, the patterns contained in two registers can be compared digit by digit by logic elements and the result registered as another pattern of digits. The resultant pattern might indicate that the two original patterns are identical, or that one is numerically larger than the other; it might likewise be arranged that the resultant represents the numerical difference between the contents of the two registers, or their numerical sum.

As with electronics, so also a minimal knowledge of logic and logic elements is required for the understanding of computers and microprocessors. For the reader who lacks this knowledge, appendix B should provide an adequate introduction.

Highways or Buses

Data patterns being transmitted from register to register generally require as many wires or conducting paths as there are bits in a pattern or word. Due to the limitation on the number of terminals to a chip (pin-out) there cannot be many such sets of conducting paths. Thus, particularly in microprocessors, when several registers need to be interconnected, this is done via a common highway or *bus*. The output of one register is gated on to the bus at the same time as is the input of the receiving register. The word bus comes from the Latin *omnibus*, meaning 'for all (things)'—it has become abbreviated by electrical engineers just as it has for public transport users. An 8-bit microprocessor depends heavily on its 8-bit wide, or 8-conductor, data bus to which all its working registers are connected. Necessarily such a system entails the time-sharing of the bus if several transfers are required between different pairs of registers, and this has to be arranged by the designer. Fairly simple data manipulation operations may require a succession of time slots to be allocated to these transfers within the device and these are controlled within the microprocessor by sets of the operations called microinstructions, forming the microroutines or microprograms. Generally, although these are transparent (or invisible) to the user, they do become apparent when dealing with timing and speed considerations, since the processor may take several 'clock' pulses to perform an operation. The term 'clock' is used for the basic timing pulses which the machine needs to synchronise its operations. For most microprocessors the clock pulse rate is one or two million pulses per second.

Microprocessor Configurations

In this book we are mainly concerned with 8-bit microprocessors of the most conventional kind. These have the entire CPU on one chip with separate chips for memory stores of various kinds and ancillary units for peripheral handling. Commonly the whole assembly is marketed as a single entity, mounted on a printed circuit board. As a very rough guide, the whole device costs ten to twenty times the quoted cost of the microprocessor chip itself. It is normally equipped with a ROM which contains a set of utility programs for operating and monitoring the processor. Microprocessors are designed primarily to be used in dedicated roles: these system assemblies are intended to help the designer to make a start in developing and testing the dedicated systems.

An important and emerging configuration is the single chip microprocessor system. These have the CPU, storage and peripheral controllers all on the one chip and are of particular use for highly specialised roles such as pocket calculators. At present most of these devices have to be programmed during manufacture and require a somewhat complicated design approach. Devices are now appearing, however, for use in much the same kind of role but which are programmable by the user by a fairly simple technique. These single-chip devices will be described later in this book.

A third category of devices will only be mentioned; these are processor chips which either work on 16-bit data or are of what is called 'bit-slice' construction. The former are especially suitable to form the processing units of 16-bit minicomputers and are used as such. The bit-slice devices are an arrangement in which the processor is devided up into its major organs each on a chip and can be put together to suit the designer's requirements for minicomputer or even large array processing computer systems. They are beyond the scope of this book but ample descriptive literature for them is available from the manufacturers.

2 A Simple Central Processing Unit

Everything in a computer is there for a purpose. Another way of expressing this is to say that every feature or detail exists as the logical solution to a design problem. The same problems are often solved by different designers in different ways, but generally the simpler the problem the simpler the solution. For this reason the easiest approach to understanding computers is by first defining and understanding the design problems, preferably starting with the simplest. As we become more aware of the nature and details of the problems we can delve more deeply into the details of the solutions and the process can be carried out by easy stages. This is a good maxim for the study of anything that is complicated.

First let us consider the arrangement and operation of a really simple CPU and how it could be implemented. As the limitations of the simple arrangement become apparent we can then consider the necessary additions and modifications to it. Figure 2.1 shows a schematic diagram of a simple CPU. The arrangement shown is quite general. It is equally applicable to computers, data processors, minicomputers and microprocessors, great or small, in the current state of the art.

The first task of the CPU is to find and fetch the instruction to be obeyed. It must then decode and execute it.

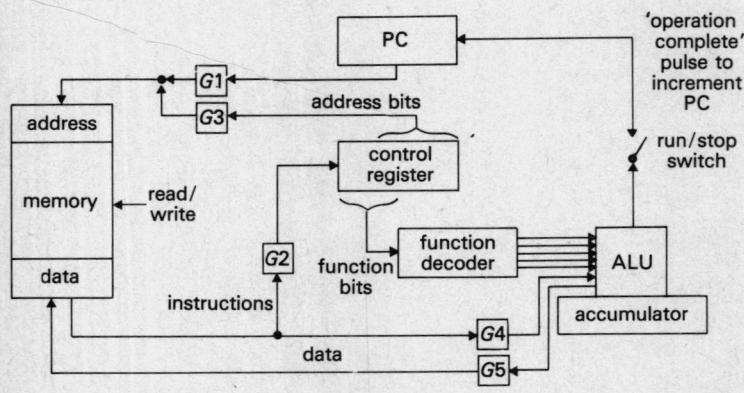

Figure 2.1

A Simple Central Processing Unit

In modern computer practice the instructions forming a program are listed consecutively and stored in consecutive locations in memory. Thus to start a routine of instructions the location or address in memory of the first is loaded into the register shown as the *program counter* (PC). The *fetch instruction* phase then becomes

(1) Transfer the value of the PC contents to the store address port and send a 'read' signal to the store
(2) Transfer the value which appears at the store data port to the control register. This concludes the fetch instruction phase.

In hardware terms these operations mean that during a particular clock pulse time the gates $G1$ are opened between the PC and address port and the 'read' signal is sent to store on the line marked read/write. Then after the store has had time to operate, the gates $G2$ are opened and the data, that is, the instruction required, is pulsed (or loaded) into the control register.

The *execute* phase depends on whether the instruction fetched calls merely for an operation or is of the memory reference type. In either case it is started by the decoding of the pattern of bits which describe the instruction.

The control register is divided into two fields, the function field and the operand address field. In the first place, the bits which comprise the function field are passed immediately to the control logic unit which decodes them. If they are decoded as an operation instruction the control logic unit causes the required operation to be performed. In this case the execute phase is then also complete.

If the function is decoded as a memory reference instruction the following sequence takes place

(1) the gates marked $G3$ are opened so that the bit pattern in the address field of the instruction is transferred to the store address port and 'read' is signalled to the store on the read/write line.
(2) when the store has had time to operate, the data pattern (that is, the operand) is transferred from the store data port to the ALU, through the gates marked $G4$.
(3) the ALU executes the function required as instructed by the control logic unit. When this is finished the execute phase has been completed.

Consideration has been given here only to instructions requiring data *from* store. Another essential instruction calls for data to be sent from the ALU and written into store. In this case the sequence is

(1) the address named in the instruction is sent to the store and the data in the ALU is sent to the store data port, through gates $G5$
(2) 'write' is sent to store on the read/write line and the store takes in the data and places it in the desired location.

The phase is complete when the store has had time to perform its write function.

From a hardware point of view these sequences merely entail the opening of the appropriate gates during the clock pulse times which allow the data transfers to take place in the proper order. Generally microprocessors use microcircuit storage which means that the store operation is so quick that the CPU does not need to wait for the store to read out or take in data. However, the faster the store, the more it costs. It is often necessary to use slower store elements for economic reasons and in this case timing problems occur and need to be dealt with.

To make the process continuous, that is, for the machine to work through a sequence of instructions, on the completion of the execute phase the control logic unit causes the program counter to be incremented so that it contains (points to) the address of the next instruction. This whole cycle, as described, comprises the execution of one *machine instruction*.

Before dissecting the sequence of instruction execution in greater detail it is as well to consider some other aspects. Figure 2.1 shows a register attached to the ALU, called the *accumulator*. The term is used in computers in much the same sense as it is used by the betting community: it accumulates totals. Logically an instruction for a calculation needs the definition of two operands and a place for the resultant. An addition instruction written in full would be, 'add the contents of location A to the contents of location B and place the resultant in location C'. Written in one form of abbreviated notation used by computer people this could be written

$$C(C) = C(A) + C(B)$$

The shorter the instruction the more easily and economically it can be stored or transferred. The designers of many machines have therefore resorted to what is called 'single address instruction' format. Using this arrangement all operations are performed on the accumulator register or its contents. The addition instruction can then be written, 'add the contents of location A to the contents of the accumulator, leaving the result in the accumulator'. It can be shortened further to, 'add A' since only A, the address of the stored operand, need be explicitly stated. The references to the accumulator are implicit in the function and need not be restated. The abbreviated notation becomes

$$C(\text{Acc}) = C(\text{Acc}) + C(A)$$

meaning, 'replace the contents of the accumulator with the sum of its present contents and the contents of location A'. Abbreviations and abbreviated notations do not necessarily make life easier, but for a designer to be able to abbreviate an instruction word certainly makes his life easier.

In many machines designers have found it desirable to have more than one accumulator or to have several central registers, one or more of which can be used as accumulators. In this case the function part of the instruction must define the particular registers to be used, thus requiring it to be longer, that is, contain more bits. In machines which have a fairly short word length, this poses problems. It particularly poses problems in machines having 12 or 16-bit word lengths, like most minicomputers since it leaves only a few bits for the address field.

The designers of 8-bit microprocessors realised that they were faced with an impossible problem from the outset and so they developed a stratagem that distinguishes these machines from the larger ones. In 8-bit microprocessor usage a memory reference instruction is written as 2 or 3 bytes or words held in consecutive store locations. Where this kind of addressing is used the PC first points to the first byte of the instruction which contains the function bits, and this is loaded into the control register which then only needs to be 8 bits long. The control logic detects that the instruction is of the memory reference kind and merely increments the PC to point to and access the operand. A sad consequence of this is that the PC no longer merely points to consecutive instructions as it does in larger computers. This makes the program listing more complicated and program debugging more difficult. A particular difficulty that arises from this is in the working out of 'jump' instructions.

Jumps

The simple CPU as described above would merely run through a sequence of instructions. Probably the greatest contributory factor to the power of computers or microprocessors is that they can make program branches, or *jumps*, automatically as the result of decisions that they themselves can make on logical premises.

A program branch is an arrangement by which the processor can make a test on data within it, or supplied to it, and depending on the result of the test, it can either continue in its current sequence or branch into another. A number of instances of such branching would occur in the sorting operation described in chapter 1. After performing the comparison of each pair of numbers and listing the larger, the machine would jump back (loop) to the beginning of the comparison routine and repeat it with new data. After reaching the end of each pass it would start the process on the new list. It would check for swops and when it had completed a pass with no swops, it would branch or jump into the output routine. The premises on which each decision would be based would be of the simplest kind such as, if the contents of the accumulator represents a negative number then branch, otherwise continue with the current sequence'. The premises must be contrived by the programmer so that the machine merely implements his previously ordained decisions.

The kind of branching described above is called 'conditional' branching. A simpler kind is unconditional branching. The instruction merely says, 'jump to N' or 'go to the instruction in location N'.

The hardware implementation for branching is quite simple. The form of the instruction is just like any other memory reference instruction. When the function part is decoded the address bits need to be sent to the PC instead of to the store. This entails the provision of another data path not included in the simple CPU illustrated in figure 2.1. As data paths proliferate the 'simple' CPU begins to assume the complexity of a knitting pattern. For instance, we have to consider data paths which interconnect the CPU with peripherals, in particular the control console. Even minicomputers normally display the contents of at least the accumulator, the program counter and control register. There also need to be data paths for control switches and the like. Now the CPU of a microprocessor has to have data paths from itself to the store address and data ports which generally total $16 + 8 = 24$. When power supplies, interrupt lines, status indicators and clock inputs are also considered it is easy to see that a pin-out of 32 or even 40 terminals does not allow the microprocessor luxurious provisions. Considerable ingenuity has to be used to get round this constraint and it poses an intriguing problem for the system designer.

In 1951 M. V. Wilkes* postulated an arrangement for a CPU in which all the registers and processing units were connected through gates to a common highway. Access to and from the highway was regulated by microinstructions which controlled the various gates: the set of microinstructions which went together to implement a machine instruction formed a microroutine; and the collection of these that provided the machine's instruction set formed a microprogram. The great advantage which this arrangement conferred was that it gave total flexibility to the designer even though it was at the cost of speed of operation. There are a great number of machine instructions that could be desired but, generally, only a few can be provided. The selection of which few poses a problem since it depends very much on the tasks the computer is expected to perform and on the preferences of the user. The microprogrammable computer allows the same basic hardware structure to remain unchanged while the instruction set can be varied to suit different requirements merely by altering the microprogram. A number of machines have since made use of this facility and the technique has become well established and highly developed.

---

* The best way to design a computing system, *Proceedings of the Manchester University Conference on Computers* (Ferranti Ltd)

A Simple Central Processing Unit

The fabrication process by which a microcircuit is made and by which a microprocessor CPU is manufactured on a chip makes it highly undesirable to have any number of data paths criss-crossing. This, and the pin-out, constrain the designer very strongly towards a single highway configuration. Also the process of fabrication makes it much easier to produce arrays of storage elements rather than logic networks; a microprogram store rather than a complex logic network therefore offers a considerable attraction. Most microprocessor CPUs are therefore manufactured in this form.

Let us therefore now reconsider our simple CPU laid out in conformity with Wilkes' microprogrammable computer concept. Figure 2.2 illustrates such an arrangement, while figure 2.3 illustrates how the

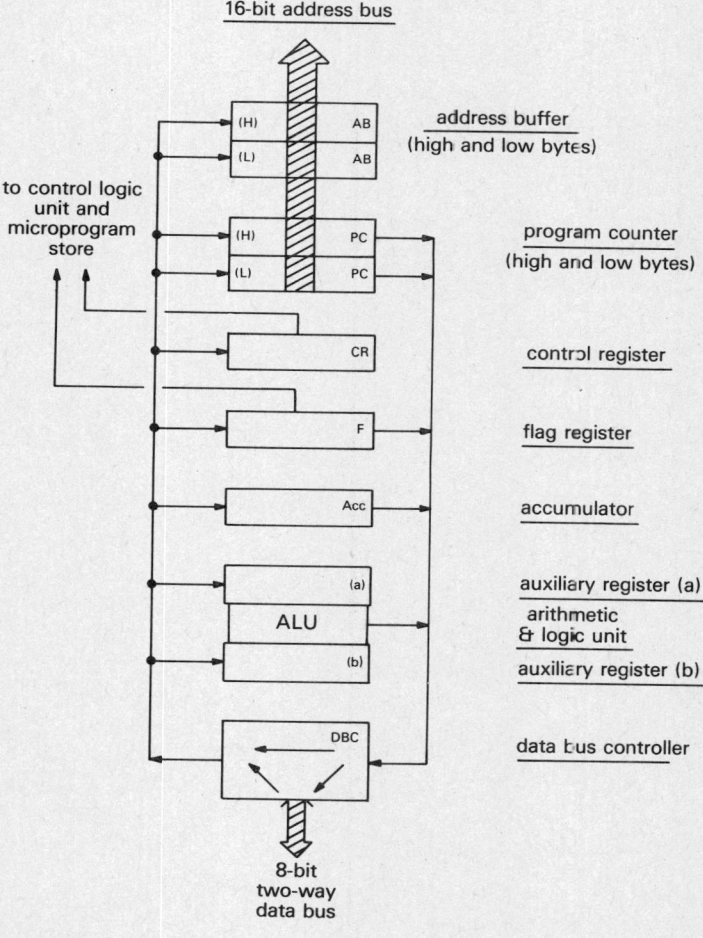

Figure 2.2

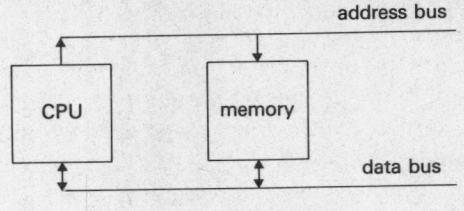

Figure 2.3

CPU is connected to its memory store. The registers shown and labelled in figure 2.2 are each of 8-bit length. Now the total number of patterns which can be formed by 8 bits is $2^8 = 256$ and therefore an 8 bit register can only identify or indicate 256 numbers. Even the simplest program for practical use will require many more storage locations than this. The address bus therefore is 2 bytes wide and the address buffer and program counter each comprise a high byte and a low byte. This allows for an addressable store of up to $2^8 \times 2^8$ locations, that is, 65 536. It is easier in dealing with computer stores, which are numerated in binary notation, to refer to this value as 64 k bytes where k = 1024 or 2^{10}. Likewise it is convenient to refer to a block of 256 store locations which can be addressed by a single 8-bit byte as a page. Thus, 16 address lines (or bits) can be used to address 256 pages of storage, as an upper limit. It is very rare indeed for a microprocessor in a conventional role to be furnished with such a large store, but the large address capability provides many other useful facilities.

The input/output data port (the data bus controller of figure 2.2) is an arrangement of gates so that it can close the CPU from the external data bus and make its internal data bus continuous. It can also connect the right-hand highway on to the external data bus to put data out and it can also connect the external data bus on to the left hand internal highway.

The ALU is shown attached to two auxiliary registers which provide the only input access to it. It has an output on to the right-hand highway on to which it can place its resultants.

No gates are shown on the diagram for the sake of simplicity. Each register has gates both on its inputs and outputs and these are controlled by the control logic microprogram.

The use of the register F will become apparent as we proceed. Conditional jumps have been mentioned already and one use of this register is to hold the result of tests on which conditional jumps depend. It is for this and similar reasons that this register, like the control register, is shown as having an output data path to the control logic unit.

There is one more rule of the game: the registers shown are simply registers, no more and no less. They cannot count or shift or do anything except hold data loaded into them. Any data manipulation has to be performed by the ALU.

Finally we should note that since the addressing capability, and the actual addresses sent to the store address port, are now 2 bytes long, there will be problems for an 8-bit machine in fetching data from any store location other than that immediately following the instruction function byte as described earlier. This form of operand addressing is called *immediate addressing*. There are several others and these will form a topic for later consideration in detail.

MICROPROGRAMMING

Even if the reader has no serious intention of ever designing a CPU, micro or otherwise, it is a salutary exercise to work through the detail of one or two machine instructions at the microprogram level. As examples let us take the following two machine instructions: 'load immediate', that is, load the accumulator with the contents of the next location, and 'add (m,n)', that is, add the contents of location m,n to the contents of the accumulator. Let us consider these two instructions as occupying consecutive store locations starting from p,q. We assume m,n,p,q to be 8-bit binary numbers. We have then

Location	Contents
p,q	load immediate
$p,q + 1$	operand of load immediate
$p,q + 2$	add (m,n)
$p,q + 3$	m ⎫ high and low bytes of address
$p,q + 4$	n ⎭ of operand of add instruction

The result of this operation should be that the accumulator ends up containing the sum of the values held in locations $p,q + 1$ and m,n.

Assume the value p,q has already been set into the PC. Let us also use letter symbols to designate the registers for the sake of brevity. Assume also that the PC and the address buffer can be gated directly on to the address bus, as is the normal arrangement.

The process then takes place as follows

(1) $C(PC) \rightarrow$ address bus: 'read' signal to store
(2) Store data $\rightarrow$ CR (via data bus controller)

This concludes the fetch instruction phase for the first instruction

(3) $C(CR) \rightarrow$ control logic unit
(4) $C(PC) \rightarrow (a)$: 'increment' to ALU ⎫
(5) ALU resultant $\rightarrow$ PC ⎭ Increment program counter
(6) $C(PC) \rightarrow$ address bus: 'read' signal to store
(7) Store data $\rightarrow A$ (operand loaded into accumulator via data bus controller)

This completes the 'load immediate' instruction

(8) $C(PC) \rightarrow (a)$: 'increment' to ALU ⎫
(9) ALU resultant $\rightarrow$ PC   ⎬ Increment program counter
(10) $C(PC) \rightarrow$ address bus: 'read' signal to store
(11) Store data $\rightarrow$ CR (via data bus controller)

This concludes the fetch instruction phase for the second instruction

(12) $C(CR) \rightarrow$ control logic unit
(13) $C(PC) \rightarrow (a)$: 'increment' to ALU
(14) ALU resultant $\rightarrow$ PC (value now $p,q + 3$)
(15) $C(PC) \rightarrow$ address bus: 'read' signal to store
(16) Store data $\rightarrow$ address buffer, high [buffer now holds $(m,-)$]
(17) $C(PC) \rightarrow (a)$: 'increment' to ALU
(18) ALU resultant $\rightarrow$ PC (value now $p,q + 4$)
(19) $C(PC) \rightarrow$ address bus: 'read' signal to store
(20) Store data $\rightarrow$ address buffer, low [buffer now holds (m,n)]
(21) Address buffer $\rightarrow$ address bus: 'read' signal to store
(22) Store data $\rightarrow (a)$
(23) $C(A) \rightarrow (b)$: 'add' to ALU
(24) ALU resultant $\rightarrow$ A [A now holds $C(p,q + 1) + C(m,n)$]
(25) $C(PC) \rightarrow (a)$: 'increment' to ALU
(26) ALU resultant to PC

Which completes the process and leaves the CPU ready to fetch the next instruction.

The reader may well have noticed that, in the listing, the process of incrementing the PC has been abbreviated: the PC has a high and low byte. The full incrementing procedure would be first to increment the low byte, as in the list, and then to increment the high byte if the carry flag in the F register has been set by the previous addition. In some microprocessors this latter step is omitted and it is left to the programmer to increment the high byte when a page boundary is crossed. It provides, naturally, ample scope for errors of omission by the unwary.

In order to increase speed and gain efficiency the designer would arrange for as many as possible of these operations to be performed simultaneously. For instance, since the store outputs data immediately on demand (1) and (2) could be performed at the same time, as could (6) and (7), and all the other store accesses. With fast logic (3) and (4) could doubtless be performed within the same clock pulse time as could (12) and (13) and other decoded operations. The shift of the accumulator contents to the auxiliary register (b) would almost certainly be contrived to take place simultaneously with some other activity. Even so, a lot of things have to happen to implement even the simplest machine instruction; the time-sharing of the common data highway slows down the machine's operation and entails several clock pulses for each instruction. Design for speed entails extra highways and extra logic units; the designer has to decide which constraints he must accept and which complexities he can afford. Some processors for instance have separate incrementation arithmetic logic dedicated to the PC; the more complexity that is designed into a chip the more it is likely to cost either through increased size, or through lower yield.

THE ARITHMETIC AND LOGIC UNIT (ALU)

We have considered the management and data routing functions of the simple CPU. We should now consider the arithmetic and logic unit since this, after all, is the part that does the business. All the other activities exist only to push the right data through the ALU at the right time. The details of the logical anatomy of the ALU are properly the study of students of logic design. Here we shall consider only what the ALU does and not go into much detail as to how it does it.

The ALU of a microprocessor is necessarily the simplest. The ALUs of big computers cost many thousands of pounds and are capable of very high speed operation on floating point numbers with great precision. As yet, not much of this kind of complexity can be designed on to a single chip. Microprocessors can only perform that kind of calculation very slowly and laboriously.

The following is a list of the functions that can be reasonably expected from a microprocessor.

> *Load:*—this means place a number (pattern) into a register.
> *Add:*—arithmetic addition of an operand to a register contents, producing the sum of the two. The function 'load' can be produced by adding an operand to a register already set to zero.
> *Subtract:*—arithmetic subtraction of the operand from the contents a designated register. Usually, as in 'add', the designated register is the accumulator, in which case the designation is implicit; microprocessors, though, tend to have several central registers and one of these may be explicitly designated.
> *Store:*—transfer the pattern in the accumulator or designated register to the given location in store. An abbreviated instruction of this kind could be 'STO *B m,n*' meaning, 'transfer the contents of register *B* to location *m,n* in memory'.
> *AND:*—this is a logical instruction meaning perform the logical AND operation on each bit of the contents of a designated register with the corresponding bit in the operand. It is sometimes called 'mask'; an illustration of its operation would be:

```
Register contents    1 0 1 1 0 1 1 1
Operand              0 0 0 0 1 1 1 1
Resultant            0 0 0 0 0 1 1 1
```

The resultant would be left in the designated register. It is a useful technique for selecting bit patterns. In the example we have selected the right-hand 4 bits, which in numerical notation would be the 4 'less significant' bits.

These five functions are normally the only store reference instructions performed by the ALU itself. However there are several operation instructions that an ALU can be expected to perform.

Increment:—add 1 (in the least significant digit position) to the contents of the designated register
Decrement:—subtract 1, as above
Negate or complement:—Complement each bit of the contents of the designated register, that is, set all 1s to 0 and all 0s to 1.
Left shift—arithmetic:—shift the contents of the designated register one place left. The least significant bit position becomes zero while the most significant bit is lost or transferred to the flag register in a predefined bit position.
Right shift—arithmetic:—as for left shift but the left-most bit position, the most significant, usually remains unchanged. The bit shifted out of the least significant end is lost.
Clear:—set the designated register to all 0s. This function should rarely be required if there is a load function.

This list comprises some eleven functions which could be expected to be in the repertoire of a microprocessor ALU. The ALU may be involved in a number of other functions under control of the microprogram, such as incrementation. Its activities are also monitored by the flag or machine status register, that is, flags are set in the flag register by certain things that take place in the ALU, such as carry out of the most significant digit position.

The instruction set of most microprocessors looks somewhat formidable when listed. This arises from the fact that the basic instruction set, similar to that above, becomes multiplied by the number of designated registers. For instance, if there are 12 registers that can be loaded then there will apparently be 12 load instructions or variants of the basic instruction. There are also a number of addressing strategies as will be shown later in this book. If the 12 load variants can each be used with 5 different addressing techniques, then there can be 60 variants of the load instruction. The microprocessor designer has an 8-bit byte available for function definition. There is no need to economise on the number of functions on that count. However it is important to note that most function codes do have a structure. Commonly the basic function is expressed first, then the appropriate register and then the addressing technique. The actual vocabulary with which the programmer needs to become familiar is much briefer than it seems at first sight. Usually the function list is written as a set of abbreviations or mnemonics. This makes for brevity, but these mnemonics can also be translated by a program called an *assembler*. The programmer lists his program in mnemonic form and the assembler assembles it as binary machine code. This will be discussed in greater detail in chapter 5 on progamming.

At the end of this book there is an appendix on binary arithmetic (appendix A). The ALU is entirely a binary device and does all its calculation in binary arithmetic or in a binary representation of decimal arithmetic known as binary coded decimal (BCD). It is not necessary at this stage to get involved in binary arithmetic to any depth; we are only concerned with the fact that the machine manipulates symbols, which is what, after all, most of us do when we perform arithmetic. There are a few points, however, that should be emphasised at this stage. The first concerns signed numbers, that is, positive or negative values. It is normal practice in digital computers of all kinds to work in a form of binary notation that is called 'positive number representation by binary fraction with negative numbers represented by their complement, modulo 2'; this is usually shortened to 'two's complement', although it is better expressed as 'complement-modulo-2'.

The use of complement modulo 2 number representation entails all numbers being scaled down to be fractions, which is something to which a programmer easily becomes accustomed. The importance of the technique is that all numbers represented in this form carry their sign as the most significant bit of their binary pattern: positive numbers always have a 0 most significant digit, negative numbers have a 1. If the rules are strictly obeyed the number system takes care of the sign aspect of calculation without any further intervention. For instance, if a larger positive number is subtracted from a smaller one the resultant number will have a 1 in the most significant digit position, indicating that it is negative. Thus the ALU, and the microprocessor can be made to take account of the sign of numbers; to test the accumulator for negative content merely calls for the most significant bit of the accumulator to be checked. More importantly in the implementation of decision operations the state of this left-most bit can be used as a logic level to enable a gate.

The word length of 8 bits restricts the precision of arithmetic done directly in the ALU. Note, particularly, that multiply and divide do not feature in the list of ALU functions spelt out earlier. Two 8-bit numbers multiplied give a 16-bit resultant, requiring double length register operations. Multiplications leading to products of only 8 bits are trivial. Higher precision binary arithmetic is quite possible in 8-bit machines by performing fairly simple routines and using two or more registers to hold numbers. It is here, typically, that the flag register becomes important since it holds bits carried out of the most significant position of the ALU resultant, sign bits, overflow bits and similar indicators which are useful in multiregister arithmetic.

Normally the pure binary arithmetic operations are used mainly in addressing operations and store management, which lend themselves to binary arithmetic since they are enumerated in binary. For higher precision calculations it is normal to work in BCD mode, with decimal digits represented by their binary equivalents. Calculations can then be handled quite efficiently by treating each number as a string of symbols and using look-up tables to aid in multiplication. Division is always a problem in any computer and this is particularly so with the microprocessor. Division routines are always long and slow in operation. A useful strategem is to provide a fast routine for developing the reciprocal of the divisor and then multiplying. Numbers represented in BCD require 4 bits for each decimal digit. Thus they may be held 1 digit to a word, which makes for simplicity, or 2 digits, which makes for speed. This latter form is called 'packed BCD' and most microprocessors are equipped to deal with it. The ALU is essentially a pure binary device; if however two packed BCD bytes are added in the binary ALU, only quite a simple logic arrangement is required to adjust the result into packed BCD form. Thus most 8-bit microprocessors are equipped with a useful ALU function called 'decimal adjust'. This detects the need for the adjustment and executes it. It also, if necessary, sets a flag in the flag register called 'half carry', that is, it sets a 1 in the designated bit position if there is a carry generated by the decimal adjust operation between the lesser digit and the greater, that is, between bit positions 4 and 5 in the ALU. This bit can be tested by program and facilitates BCD operations.

We do not need to become deeply involved in the techniques of number juggling to understand the operation of microprocessors. We do need to understand these techniques very thoroughly, though, if we become seriously involved in machine code programming and writing arithmetic routines. We must always remember that we are dealing with a machine and not an intelligent being. It will do nothing that it is not told to do but will also do stupid things if it is told to.

3 The Memory or Store

Memory storage devices lend themselves particularly well to the techniques of microcircuit fabrication since they consist primarily of regular arrays of identical elements. The technique of manufacturing them commercially preceded that of the microprocessor itself. The necessity of expanding the market for these storage devices lent a good deal of impetus to the development of the much more difficult and less lucrative technology of microprocessor production. Microcircuit storage devices also have a significant effect on the whole nature of the microprocessor system. Designers of computer systems before the days of the microprocessor were continually confronted with the major problem that there could never be enough fast working store. However big the system, the tasks for which it was designed always demanded more fast store than the customer, even the richest, could afford. Hence computer stores were always hierarchic with as much fast store as could reasonably be provided backed up by mass stores for which the cost per bit was at least an order of magnitude less.

The most important and successful form of computer working storage was based on the use of ferrite cores. These are tiny rings of sintered magnetic material, threaded on fine wires by a tedious and exacting manual process. Core stores were, and still are, expensive as a result, but they have one outstanding merit in that they are non-volatile. What this means, in less technical terms, is that when power to the store is switched off the data held in memory is safely retained. So long as the correct switching off procedures are observed the data contained at switch off remains uncorrupted for an indefinite period. This allows programs of a semi-permanent nature, such as operating systems, to remain safely within the store on switch off or power down; however when core stores are in operation they are quite fast random access stores.

Microelectronic random access stores are cheap. They are so cheap that a microprocessor can usually be furnished with all the random access store it needs. However they are volatile. On power down their contents become totally erased or corrupted. In order to keep service programs resident in microprocessor systems when they are switched off, resort has to be made to another kind of microelectronic storage device, the ROM. This is indeed non-volatile; in fact it is permanent and cannot be changed or altered once it has been fabricated.

The applications of microcomputers and conventional microprocessors are significantly different because of the significant differences in appropriate storage techniques. Essentially mini and larger computers can operate with continually changing programs; systems based on them can be improved and updated day by day. They can be used for a multiplicity of tasks. System changes can be dealt with by program modification and adjustment. For minicomputers in on-line and real-time control applications this can be very important. The system can be tuned by program changes to get the best performance under changing conditions.

The prime use of the conventional microprocessor is in a dedicated role. It is basically very cheap and therefore there is no economic compulsion to use it as a general purpose device. Its program must be developed, tested and proved and then written (or 'blown') into a ROM; once set up it should then function indefinitely without modification. If the system does need to be modified in any substantial way then a new program must be developed and written into a new ROM to replace the old one.

Nowadays ROMs are available that can be erased and rewritten but the process is not very simple. Generally these too would be used for dedicated systems, but in this case for systems produced in small numbers and not mass produced.

The problems of developing and debugging microprocessor programs means that special techniques are required for developing microprocessor systems. These will be dealt with later when the more exact nature of the problem has become more apparent.

CHARACTERISTICS OF MICROCIRCUIT STORES

A microcircuit store amounts to an array of registers. This is one of the attractions of this kind of storage and it is the reason that large and fast machines often use microcircuit stores as part of their working store, in spite of their being volatile.

There are a variety of microcircuit stores, each having its own characteristics, merits and demerits.

Random Access Memories (RAMs)

The bit cells that make up these stores are tiny bistable elements fabricated in densely packed arrays on silicon chips in the same way as microprocessors. Naturally, the simpler the cells and the smaller they are,

The Memory or Store

the more densely they can be packed and the cheaper the cost per bit.

There are two main categories of RAM: static and dynamic. The individual cells of the static RAM are more or less conventional bistables which can be set and reset like logic element bistables. They are static in the sense that they retain their state, set or reset, so long as the power supply is maintained. Like logic element bistables, they depend on amplification with one side in the 'on' condition holding the other 'off' or vice versa. The dynamic bistable cell is much simpler and depends for the retention of its state on the retention of small electric charges which leak away. In order to retain their states the bistables have to be refreshed, that is, reset and set again at short intervals. The whole array of bistables on a dynamic RAM store chip needs to be refreshed at least once every one or two milliseconds to retain its data pattern. This refresh cycle has to be arranged by the system designer. Some microprocessors are themselves arranged to control and operate this cycle automatically when connected to dynamic stores. Arrangement of the refresh cycle is not difficult and it can usually be contrived so that it takes place between operational data accesses, that is, while the microprocessor is operating internally. If this is the case then there is no loss of system speed due to the refresh requirement. The advantages of the dynamic RAM compared with static RAM, are its cheapness, speed and low power consumption. Generally dynamic RAM is more suitable for systems with large stores when the refresh overhead can be shared. Conversely static RAM is generally more suitable for small systems. Static and dynamic RAM stores are both volatile, and are more suitable therefore for storing data rather than programs.

The organisation of RAM storage on a chip may take several forms. Generally the most economical is the type that stores a single bit in each addressable location. Using these, an 8-bit store is formed of eight of these devices in parallel, each supplying 1 bit of the addressed word. A common capacity per chip is 1 k bits or 1024 bits. This number of bit cells can be arranged as a square matrix of 32 rows and 32 columns. Each row and column can then be addressed by a 5-bit pattern, and an individual cell by 10 bits. Since there need only be 1 bit of data to be read or written, the total pin-out required for data and addressing is only eleven terminals. In addition, there needs to be a terminal to receive the read/write command, the necessary power supply terminals and a further one, called 'chip enable', which as its name implies can bring the chip into, or hold it out of operation. Most manufacturers also produce 1024-bit chips configured to operate as 256×4 bits or even 128×8 bits, at a somewhat higher cost since they are more complicated and likely to be used in smaller numbers. More recently, 4 k bit chips have become generally available as well as dynamic RAM chips of 16 k-bits. There are already experimental 64 k bit chips in pilot production.

Read-only Memories (ROMs)

Like RAM stores these devices consist of a matrix, usually square, of bit cells. Unlike RAM they are commonly organised as 8-bit devices. The bit cells or bistables are set to their required values by a masking process during manufacture. This is only economical on large production runs and these devices are therefore either dedicated to a mass production application like pocket calculators or electronic games or else to a body of service programs to form a general purpose operating system and monitor, to facilitate the working of a microprocessor. Practically all microprocessor manufacturers market these devices as part of their standard range. Today there is an increasing number of factory made ROMs which carry utility programs, and BASIC interpreters. Doubtless it will not be long before many more 'library programs' become sufficiently standardised for masking into ROMs. Programs written in ROM are often called 'firmware', implying their mid-position between software and hardware.

Programmable Read-only Memory (PROM)

In order to make it possible for individual users to make single or small quantity ROMs, devices are fabricated which can be written, or 'blown', by the user. The early types of these were masked to 1 in all bit positions but the masking conductors were made fusible. What this means is that the linkages were made of very thin tracks of high resistance or low melting point metal. It was then possible for the user to select current paths through the unwanted 'fusible links' and to 'blow' them by passing sufficiently high current pulses through them so that they vaporised or decomposed. It was a somewhat fallible method. Occasionally too high currents blew nearby wanted links and sometimes after a period fused links restored themselves causing 0s to become 1s again. Today by the use of better fusible material these problems have been solved so that failures are rare indeed. A particular advantage of these devices is that they need not all be programmed simultaneously, that is, programming can be done step by step as new routines are developed and tested. Once a bit or word is blown however, it cannot be modified further. To obviate this inconvenience a more advanced device is now available.

Erasable or Reprogrammable Read-only Memory (EPROM)

These devices are characterised by a quartz window in the encapsulation, exposing the chip to light. The user can actually see the pattern of the deposition on the chip surface. The technique most frequently used is called 'silicon gate': small segments or areas of the silicon deposition are isolated from the main substrate by very thin layers of silicon oxide which is a very good insulator indeed. The chip is so designed that the user can pass a series of pulses through selected segments building up charges in them. These charges remain in position for ten or more years because of the oxide layer. Even so, their potentials control the gates of the bistable elements and the device becomes a read-only memory. (The life figure of ten years is only an estimate. Perhaps after ten years of practical operation they will be found to have a much longer life.)

In order to erase data from the memory the chip is subjected to ultraviolet light for 20 to 30 minutes. The light photons cause conduction to take place, releasing the stored charges. The process of writing can then be repeated and the device becomes a ROM once more.

The writing process has to be carried out carefully. No segment must be allowed to overheat, and so to store the necessary amount of charge a series of a hundred or so short pulses has to be applied to each bit with a fairly long separation between each pulse. The process is best carried out using another microprocessor or minicomputer or by a specially built logic unit. The data to be written is stored in RAM. The pattern for the whole ROM is then written word by word by addressing and pulsing each word in turn and until recently the whole process was repeated about 100 times. Nowadays, EPROMs are becoming available which do not require the repetitions. This operation, without the use of a special unit or another processor, is impractically tedious and very prone to error.

An alternative process is now in early stages of development, the electrically alterable read-only memory (EAROM). These devices can be altered or rewritten merely by pulsing them with currents and voltages at closely defined levels. They are still far too slow to be considered as an alternative to RAM store but they do open the way to the idea of having a part of a computer's store organised on a 'read-mostly' basis: the computer could write into this part of the store routines and subroutines which it needed frequently or on a semi-permanent basis. As yet they are too expensive to use except for experiment or in applications where a small store of this kind could be economically justified. Such applications can be envisaged typically in communication networks where directories need continual updating.

Stand-by Supplies for RAM Stores

It is not difficult in permanent and static installations to have RAM power supplies backed up by batteries which replace supply power when power-down occurs. This prevents loss of data in the case of power failures and for short periods of shut down. It is less convenient for mobile or portable installations. There are however types of RAM stores available which have a low power consumption when in operation and a very low consumption indeed when the supply voltage is reduced to stand-by value. These can well be used with portable systems. A typical plug-in store board carrying 4 k bytes of store can be supported for a month or so by two pen-light batteries clipped on to the board. The board may be unplugged from the system and stored for up to the life of the batteries.

Even though the volatility of RAM stores can be made less of a disadvantage by these means, such arrangements do not, for some users, give sufficient confidence in their reliability. The search continues therefore for non-volatile storage media which are cheaper and less bulky than core stores.

This threat of core stores becoming non-viable because of cost, speed and bulk is not new: the makers of core stores have known it for years. The result has been that core stores get better and cheaper continually as do microcircuits. Year by year ultimate limits have been postulated and surpassed. It seems likely that the core store will be with us for some time yet.

ADDRESSING STRATEGY

An 8-bit register is capable of indicating the integers from 0 to 255. A computer memory store, to be of any practical use, needs many more locations than 256 and each must be addressable. Manifestly the addressing mechanism for an 8-bit system will require more than 1 byte to define an address. If 2 bytes are used then they can define 256×256 = 64 k or 65 536 locations: this is quite adequate, even luxurious, for most microprocessor applications.

The PC of a microprocessor is normally a 16-bit register which can signify address values to cover the whole store address field. The 8-bit size of most other registers, especially the ones used in arithmetic processing, puts constraints on operand accessing and, particularly, on jump instructions. A jump instruction must unambiguously define within the whole address field where the instruction is stored which is to be next in sequence if the jump is obeyed. Data and instructions in normal sequence are commonly in consecutive addresses but jumps are arbitrary. Typically, library subroutines and similar commonly used pieces of program may be located at one end of the store while the problem data and program are at the other. As the program is executed there will be jumps backwards and forwards between them.

Let us refer back to the instructions which we earlier dissected so tediously for microprogramming. The second instruction was ADD (m,n). The complete instruction required 3 bytes since the operand, held in page m, line n, required 2 bytes to locate it. There are likely to be many such instructions in a program. Processor designers and programmers try to reduce this addressing overhead as much as they can for the sake of store economy and speed of operation.

This addressing problem is not peculiar to 8-bit microprocessors: they share it with large mainframe computers. Although large computers often have quite long word lengths, they also have very large stores and these are usually too large to be directly addressed even by the longer word length. There are, in fact, several features of microprocessors which resemble those of larger computers rather than minis and this is perhaps the most important of them. This resemblance is also important for the microprocessor designer: designers of large systems have faced this problem for the past twenty years and the microprocessor designer has their experience to draw on.

The inadequacy of the word length is not the only reason for addressing strategies or ingenious addressing mechanisms. The nature of the work that computers are required to perform and the technique of programming give the processor designer considerable scope for improving its efficiency. As was stressed in chapter 1, computers thrive on highly repetitive tasks and operate on strings, or arrays, of data. The appreciation of this led computer designers, from the earliest days, to use the techniques of what is called address modification or indexing (for example, in 1949 Manchester University's 'enhanced' MK 1 had two modifier registers). The reasoning behind these techniques will be made more apparent in chapter 5 on programming. What modification means, in this context, is the automatic incrementation of operand addresses so that, as a process is performed repeatedly, it systematically scans through the lists of data.

Another important technique that has been found to be very powerful in use is that of indirect addressing or merely 'indirection'. What this means is that the address contained in the instruction is not that of the actual operand, that is, the data to be operated upon. The location defined by the instruction address contains another address which is the address of the operand. One advantage of this stratagem is that the programmer can write programs, or routines, with arbitrarily chosen addresses, at the time of writing, without knowing at that time where the data will actually reside. By this means routines need be written only once but can be used on different sets of data: The actual operand addresses can be written into the arbitrarily chosen indirect address locations at a later time or by another programmer.

The remainder of this chapter is devoted to the description of the more common addressing mechanisms. Generally, not all are included in the repertoire of any particular microprocessor.

Immediate Addressing

This is the name commonly given to the addressing mode in which the actual operand is stored in the location immediately following that of the instruction. This corresponds exactly to the same-named mode for computers having a longer word length. In these, the actual operand is written in the address field of the instruction. Sometimes this mode is called 'literal addressing'. The reader who is familiar with machine code programming in minis or larger computers may already have noted that the instruction word of 8-bit devices is made to correspond to that of their larger brethren by using 2 or 3 bytes, consecutively stored, to amount to the equivalent of a single word instruction of longer word-length. For example

Location	Contents
m,n	[instruction]
$m,n+1$	[operand]

Sometimes a 16-bit operand may be specified. In that case, a third consecutive location will be occupied by the less significant or 'low' byte of the operand. Naturally, the instruction decoding logic must be made to call in the consecutive operand bytes and place them in the proper places and must increment the program counter accordingly. The requirement for this must be expressed in the function code of the instruction, either implicitly or explicitly.

Direct Addressing

In many processor systems, 'page 0', the lowest 256 numbered locations of the store, is dedicated to special activities such as peripheral control and operating system functions. It is however still partially at the disposal of the programmer who may wish to store in it values of constants and such like to which he wishes to make frequent reference. The direct addressing mode allows this with the minimum of complication. When the function code of the instruction signifies to the decoding logic that direct mode is required the microprogram accesses the byte in page 0 which is addressed by the byte immediately following the instruction. In this way, the operand may be obtained using only one address-byte and the whole memory reference instruction occupies only two bytes of storage

Location	Contents	
$0,q$	[operand]	(page 0, line q)
.		
.		
.		
m,n	[instruction]	(specifying 'direct' and page 0)
$m,n+1$	[q]	(operand address, page 0)

This mode may be usefully extended to deal with 16-bit or 2-byte operands, which will be located at q and $q+1$.

Current Page Addressing

This is the converse of direct or page 0 addressing and is used more commonly in minicomputers and 12 or 16-bit micros or minis. If the function code signifies this mode to the decoding logic, the processor uses the byte which follows the instruction as the operand address but in the current page as defined by the most significant byte contained already in the PC. This mechanism again allows a complete memory reference instruction to be contained in 2 bytes

Location	Contents	
m,n	[instruction]	(page m, byte n)
$m,n+1$	[q]	(operand byte address)
m,q	[operand]	(page m, byte q)

Relative Addressing

This is another, and much used, mechanism for containing a memory reference instruction in two bytes. In this mode the function code instructs the decoding and microprogram logic to take the contents of the byte immediately following the instruction and use it as a 'displacement'. Usually the displacement is expressed as a signed binary number in the range -128 to 127 or, expressed in mathematical terms, $-128 \leq d < 128$. Consider first a positive displacement

Location	Contents	
m,n	[instruction]	(specifying relative mode)
$m,n+1$	[d]	(displacement value)
.		
.		
.		
$m,n+d$	[operand]	
.		
.		
.		
$m, 255$		(page boundary)

Usually a displacement is not permitted to transgress a page boundary and this must be carefully watched when programming. Consider now a negative displacement

Location	Contents	
$m,0$		(page boundary)
.		
.		
.		
$m, n-d$	[operand]	
.		
.		
.		
m, n	[instruction]	
$m, n+1$	[d]	(displacement, a negative number)

Again the page boundary must not be crossed. If it is, the page value, m, remains unchanged and the wrong operand will be accessed: readers familiar with binary arithmetic will be able to work out what the actual operand address will be in the case of this error. It is useful to know this for program debugging should this error have occurred.

If the processor logic is sufficiently sophisticated, then the transgression of the page boundary merely increments or decrements m, the page number, and the page boundary limitation no longer applies. It is important for a programmer on any machine to be fully aware of the particular details of this addressing mode since it can be a fertile source of error. On the other hand it is a most useful technique especially in the construction of program 'loops'. These are the programming ploys used when a routine is performed repeatedly. Each time it ends, a test is made and it jumps back to the start of the routine, so long as the test indicates that it should. Often such routines are only a few instructions in length and are repeated very many times. Obviously it is a considerable saving if such jump instructions can be contained in 2 bytes. Likewise in programming it is often easier to specify a short displacement than to have to specify within the instruction the exact address value of the destination of a jump.

Indirect Addressing

This is the mechanism described earlier in which the first operand accessed is actually the address of a second operand which is the one required for use. Generally, since the second operand can be anywhere in store, the first or primary operand must consist of 2 bytes, a high, or most significant, or page byte and a low, or least significant, or line byte.

Paradoxically the real value of indirect addressing can best be shown in the context of page 0 or direct addressing. The programmer may wish to access a particular routine or piece of data but may not yet be ready to specify its final location. In this case a useful strategem is to 'address indirect' in page 0, since this instruction only requires 2 bytes. In page 0, the 2 bytes at the address indicated by the instruction can be loaded with the second or proper operand address when it has been finally assigned. For example

Location	Contents	
0,q	[r]	(page byte of operand address)
0,$q+1$	[s]	(line byte of operand address)
.		
.		
.		
m,n	[instruction]	(function: indirect, page 0)
$m,n+1$	[q]	(page 0 address)
.		
.		
.		
r,s	[operand]	(required operand)

The combination of page 0 and indirect addressing is most useful but it is not necessary for it to be restricted to page 0. Any 2-byte operand however arrived at could well be made the address of a secondary operand using the same procedure. Not all microprocessors allow explicit indirection. We shall see later when we re-examine the structure of the CPU that the provision of indirection poses problems and adds complication. More importantly, the great advantage of indirection lies in its ability to allow programs and routines to be moved about or 'relocated'. Their assigned position at any time is not important so long as they are linked to primary operand addresses. Microprocessors, designed for dedicated roles, have their programs written indelibly into ROM and there is no possibility of relocation, even if there is a need for it.

So far we have concentrated on addressing mechanisms which can be accomplished with instructions of 2-byte length. Naturally these are desirable and should be used where possible. Generally, however, memory reference instructions are of 3 bytes. Hence the name for the next mechanism, is 'normal addressing'.

Normal Addressing

This corresponds exactly to what would be called 'normal' for any computer having a 24-bit word length. In the microprocessor world, it is often called extended addressing since it is an extended form of the direct addressing mode which was described earlier. The first byte, as usual, is the instruction byte; It is followed immediately by 2 bytes which define page and line anywhere in store to access the operand required. For example

Location	Contents	
p,l	[instruction]	
p,l+1	[m]	(high or page byte)
p,l+2	[n]	(low or line byte)
m,n	[operand]	

A large proportion of all instructions in a program would be normal address or extended address.

Address Modification

In the original Von Neumann computers all programs and data shared a common store. In 1946 Von Neumann postulated the idea that instructions themselves are patterns of binary numbers and could be operated on by normal arithmetic techniques just as though they were data. Considered generally, this poses an alarming prospect but the modification envisaged was very limited. The thinking behind Von Neumann's postulate was that, in accessing an array of data in consecutive store locations, instead of writing a separate fetch instruction for each one, it would be more efficient to write it only once. Then each time it was used, it would be passed through the arithmetic unit and have 1 added to it in the address field. In the earliest machines this is exactly what was done. It occurred to the Manchester computer designers however that they could save a whole group of instructions for performing each modification by providing a separate additional modifier register. When an instruction is marked for modification the address in the instruction, which itself remains unchanged, is added to the number held in the modifier register to give the 'effective' address. Ideally, this modifier register could have its own arithmetic facilities: for instance, it could be automatically incremented or decremented each time it was accessed. It could have flags or indicators attached to it so that it could be tested to show if it contained, for instance, all zeros.

Microprocessor designers have made considerable use of modification techniques: the more modern, and more commonly used name for them is *indexing*. It can have a number of variations and can be used in combination with direct or indirect addressing.

Indexed Addressing

In most microprocessors a 16-bit register is provided and called the index register. An indexed instruction must indicate that this register is to be used and will be followed by another byte which indicates the displacement in some way. The byte could itself represent the displacement, as in an 'address relative' but this would be of limited use. Usually the byte or bytes following the instruction indicate, or 'point to', a register or store location which contains the modifier. If a store location in main store is indicated, then this gives great flexibility but the incrementation of the modifier has to be performed by the ALU. It might seem that there is little to be gained by using this technique. We must remember however that microprocessor programs finally end up in ROM store. Thus the address in the instruction cannot itself be incremented. To modify it the contents of a RAM location can be used.

If a microprocessor is well provided with central registers it is common to use one or more of these as modifier registers. Naturally, since both RAM locations and central registers are of only 8-bit length, modifiers are restricted to numbers up to 256.

There are a number of possible ways of implementing indexing both by hardware and software. Ideas of its potential should become more apparent when we have further considered the CPU and some of the techniques of programming. This is another function that an intending programmer should study fully for the particular processor that is to be used. In fact it may even provide a criterion for choosing which processor is to be used in the light of the task to which it is to be dedicated.

Zero Addressing

This is the name that computer scientists and engineers use for what is more colloquially called 'stacking'. It occurred to computer designers of large computers that in some processes, particularly in the evaluation of mathematical expressions, it was useful to provide a mechanism called a 'stack'. For instance, consider the expression $(A + B)*(C + D)*(E + F)$, where the asterisk sign denotes multiplication. The process of evaluation entails loading A and adding B to it. This subtotal must then be stored somewhere while $(C + D)$ is evaluated. The value of $(A + B)$ could then be recalled and used as the multiplier. The result of this would then need to be stored while $(E + F)$ is evaluated. It then needs to be recalled as the multiplier. The storing of partial totals can be made less of a chore if there is one, or preferably several, spare registers handy. Even so, each needs to be addressed.

The provision of a stack makes for simplicity and efficiency. Conceptually it consists of a block of registers so arranged that if a word or byte is sent to it it rests in the top location. If a further word is sent, this pushes down the first and replaces it at the top and so on. We then have a 'push down stack'. A command to take a word off the stack extracts the top one and causes the others to pop up one place. In the evaluation of long mathematical expressions, there sometimes need to be a number of words pushed on to the stack and later popped off it again in last-in, first-out order.

The original stack implementation in the English Electric KDF9 provided for 16 stack registers directly associated with the machine's accumulator. An early microprocessor, Intel 8008, had 8 stacking registers associated with the PC.

One problem associated with small stacks is that it is not easy to ensure that they do not become overloaded for if they do the data at the bottom of the stack is lost.

The computer engineer does not see the stack in the conceptual form described: the way he implements a stack is to provide an array of numbered registers and a pointer or incrementable counting device. Thus with the stack empty, the counter points to 0; push one word on to the stack means load it into register 0 and set the counter to 1; push another and it is loaded into register 1 and the counter increments to 2. Popping is the reverse process. The register pointed to by the counter is unloaded and the counter decremented, and so on. The term 'zero addressing' refers to the fact that the stack and unstack, or pop and push, instructions need no explicit address.

The microprocessor designers, being acquainted with both the current hardware and software techniques of stacking, were quick to see what the big computer designers had seen: a stack does not need to be a fixed, dedicated set of registers. It merely needs a pointer mechanism, that is a counting register, to indicate the top of the stack or 'stack front'. It can then be incremented through an array of store locations anywhere in store and be of arbitrary size. Hence most microprocessors have at least one 16-bit register which is called a 'stack pointer'. This pointer can be set to any value to suit the programmer who can then use it to store any kind of data that can be put away on a first-in, last-out basis. A very common application is when a current program has to be interrupted or suspended. In this case all the current data of the suspended program—the PC value, accumulator contents, etc. which comprise what is called the 'status vector'—can be pushed, in sequence, on to the stack. When the program is restarted the status vector values are popped and replaced in their original registers for the program to continue from where it left off.

In high level language operation of a computer the use of addressing mechanisms is the concern of the compiler writer and they are more or less transparent or invisible to the programmer. In machine or assembly language programming they become the business of the programmer. They can, if they are properly understood and applied, be a powerful tool for good or efficient program design. On the other hand they can and do provide pitfalls for the unwary. The rules for using them are often complicated and, regrettably, often poorly and inadequately defined. The machine code programmer is advised to study them carefully and understand their implementation. He is then likely to avoid most of the pitfalls and spot the ones that he has missed fairly easily in the debugging stage. There is truly no substitute for understanding and experience.

4 Practical CPUs

The basic organisation and principles of operation of data processors and, in particular microprocessors, have been explained in chapters 1, 2 and 3. It now remains to be seen how these principles can be exploited to make useful devices.

In comparison with what we expect a microprocessor to do, the CPU instruction repertoire is tiny and the instructions themselves almost trivially simple. It can only perform sophisticated operations by using complicated routines of many simple instructions. The prime conclusion we derive from this is that it must do what it does do very quickly indeed if it is to justify its existence. A major objective of design must be operating speed.

Ultimately every process boils down to passing byte patterns through the ALU. The overall speed of operation must therefore depend, in the first place, on what is often called the 'raw' speed of the logical devices from which it is made. This is a matter of electronics and the fabrication technology involved. The design of ALUs has been a main preoccupation of computer designers since the first digital computers. By now the logic configurations have been developed to such an extent that there is little a designer can do to improve them. Thus an ALU, fabricated in a particular technology provides little scope, in itself, for improvements in speed of operation. Such speed and efficiency gains as can be made must depend on the organisation of the system.

Thinking back to the hypothetical, primitive CPU described in chapter 2, certain aspects cry out for improvement. A count of the microprogram steps involved in the execution of a simple instruction shows that the large majority are spent not in processing data in the ALU, but in fetching it and fetching the instructions. The ALU is often used for trivial activities such as incrementing the PC. This process, since the PC is 2 bytes long, also necessitates the use of the carry flag in the status register. For many useful programmed operations this flag is required to carry information forward to a following instruction and using the flag for PC incrementation would hinder this.

We note, too, that the accumulator has to be used for many things. Its contents must often be stored and then almost immediately recovered. This will become even more apparent in the consideration, later on, of programming. It would seem then that considerable advantages might accrue from the provision of some extra registers as temporary stores on the chip itself, available without all the addressing palaver.

Another strategem that should lead to faster operation is to use parallelism where possible, that is, to perform microinstructions simultaneously where they do not contend for the same facilities.

Figure 4.1 shows the register and highway layout of a considerably amplified CPU; it conforms to the Wilkes pattern as did the primitive one. Later in this chapter it will be shown that several actual microprocessor CPUs conform likewise. Mapping them on to this basic configuration is a useful approach to comparing them and analysing their characteristics.

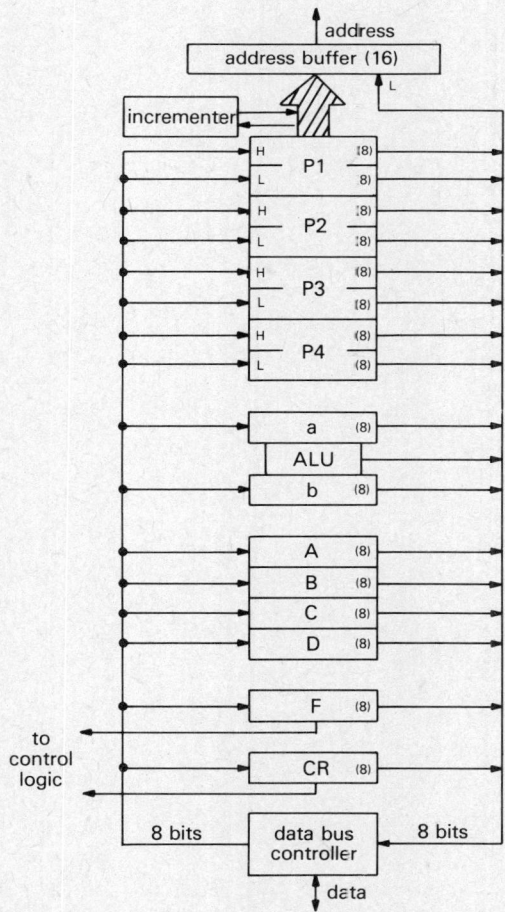

Figure 4.1

The planar fabrication technique and limitations on pin-out considerably circumscribe the options open to the chip designer as was stressed earlier. It is difficult and uneconomic to provide multiple highways. Registers, so long as they are arrays of the simplest storage elements take up little space and are comparatively easy to design and implement. Registers having built-in logic function capability are more costly and difficult. So called 'random logic' networks of arbitrary configurations of various kinds of logic elements, are still more difficult and take up an uneconomical amount of space on the limited surface of a chip. We see then that the amplified CPU of figure 4.1 is different from the primitive one mainly by the provision of extra registers.

It is convenient to consider the whole unit as being divided up into three register groups. The first group of registers at the top of the diagram will be referred to as the *pointer* group, $P1$-$P4$. They consist of a pair of 8-bit registers referred to as high and low, H and L. They are associated with the 16-bit register called the address buffer and the 16-bit wide output highway which becomes the address bus outside the CPU. This group is organised to deal with store addressing. The registers are called 'pointers' because they point to store locations.

The second group of registers A, B, C and D, are associated with the ALU and are primarily for use in dealing with the data for processing. Also associated with the ALU but connected directly, rather than through the highway, are the auxiliary registers, labelled (a) and (b).

The third group contains the control register (CR), and the status or flag register, (F). These two have their outputs connected directly to the control logic and microprogram store which are not shown on the diagram.

The Pointers

Commonly these 16-bit registers are similar in form but different in function. In a typical microprocessor configuration they might fulfil roles as follows

$P1$, the program counter   'PC'
$P2$, the stack pointer   'SP'
$P3$, $P4$, index registers   IX1 and IX2

The purpose of the PC has already been described. Suffice it to say, in recapitulation, that its task is to provide the address of the next instruction and, in immediate addressing, to be incremented to provide the address of the operand.

The nature and use of a stack has been described in chapter 3. This register can be loaded by program, as a pair of 8-bit registers, to point to the top location of the stack. Thereafter it can be incremented or decremented by a single byte instruction to 'push' data on to the stack or 'pop' it out of the stack. Pop and push are nice and brief: for larger machines the more dignified terms 'stack' and 'unstack' are usually used.

The index registers are for use in indexed addressing. Each can point to a store location anywhere in a store of up to 64 k bytes and can be loaded at the behest of the program. In larger computers, and some microprocessors, they can be incremented automatically or by an instruction, for indexing through an array. In some microprocessors instead of being directly incremented they provide a base address to which is added the contents of one of the central registers as a 'displacement'. In this case the incrementation may be performed in the central register before the addition. Thus, in operation, the contents of an index register may consist of a base address plus a displacement plus increments. It is stressed again that a programmer, before using indexed addressing, must be thoroughly acquainted with the particular technique used in the microprocessor being programmed, or the results may be dire.

An important feature of these pointer registers is that they can be incremented, usually other than by the main ALU. The question naturally arises as to how this is done. Presumably because the exact details are of little interest to the programmer the manufacturers' literature contains little or no information on this point. A notable exception is the National SC/MP, commonly referred to as 'Scamp'. Its handbook mentions that there is a 4-bit incrementation unit associated with the pointer registers. Some other manufacturers indicate the presence of some kind of incrementing device by inserting in the system diagram a box marked +1 in place of the box marked incrementer in figure 4.1, but with no other detail. Indeed it doesn't matter much to a programmer how the incrementation process is performed but it may well stir the curiosity of the person who is interested in how the microprocessor works. It is worth while here to speculate on how it may be done.

A 4-bit adder or counter unit would suffice, so long as the necessity for carry and repetition were acceptable. In the worst case, for a full 16 bits, 4 operations would be required. It is worth noting that the SC/MP normally functions with a curtailed address word, eliminating the possibility of this worst case.

The provision of an 8-bit adding or counting device would give more speed at the cost of more logic. It could well be that a designer would find such a provision cost effective.

For maximum efficiency the best but most expensive solution would be to provide at least one 16-bit register with full logical incrementation capability. Suppose one such were provided as the address buffer then the process of incrementing and data accessing could be quick and simple. The contents of the appropriate pointer would be transferred to the address buffer and incremented. The incremented value could then be placed on the address bus and, at the same time, also be sent back to the pointer register. This would require only two process steps at a cost of one fairly complex and expensive piece of logic. On the other hand, however, it could serve any number of pointer registers, all of which would need to be only of the simplest kind.

The Central Registers

In larger computers the single address instruction format is often used. If the contents of the accumulator require storage temporarily, or if a word of data is required repeatedly, the store transfer of the data can be achieved within a single instruction and little time penalty is incurred. Microprocessor store addressing is generally more time consuming and hence true single address operation is not efficient or it is slow. Resort might perhaps be made to stacking, but this is error prone and is not always convenient. In microprocessor practice it has been found more convenient to provide extra easily accessible registers to serve as temporary stores. This is the purpose of the group of central registers. In practice, one may be earmarked as the accumulator with the others merely as temporary stores; sometimes all the central registers can be accumulators.

It is necessary to have some form of indication as to which central register is to be used. Since the function part of a microprocessor instruction is necessarily 1 byte in length some of these 8 bits can be used as central register addresses. With 4 registers, 2 bits are enough to address them, leaving 6 bits for the function definition. With the limited range of ALU functions that are usually provided 6 bits are generally ample. If 8 registers were provided, then they would require 3 bits for their address indication and this could well reduce the number of function bits too much. The format in this case is really two address rather than single address but this point is of academic rather than practical interest.

Multiple central registers provide many useful facilities, most of which substantially reduce the number of store transactions that would have to be performed if they were not provided. Typical uses are to hold current count values, or frequently used constants, increment or decrement values or numbers or patterns that have to be matched or compared with strings of data, as in text handling. In reading in or outputting ASCII or BCD data each character may require masking or some similar editing process. It is very convenient if the mask byte does not have to be recovered from main store for each and every operation.

The auxiliary registers are not accessible to the programmer and are transparent as far as he or she is concerned. They are completely under the control of the microprogram or control logic and are used as temporary data stores in operations in the ALU. Like the incrementing device mentioned above they are not always explicitly mentioned in the manufacturers' literature. Generally their existence can be deduced from the operation timing since they require a program step or clock pulse to be loaded. There must be at least one auxiliary register or else it would not be possible with a single highway for 2 bytes of data to be applied to the ALU inputs simultaneously. The second auxiliary register could be somewhat of a luxury, making for simplicity at the cost of speed perhaps, but there is another reason for having two. Once they are loaded and the ALU is activated the highway is free and can be used for something else, allowing increased parallelism. If there is only one, then the accumulator or a central register must be on the highway to the ALU while it is operating.

The Control and Flag Registers

The control register holds the function byte of the current instruction. Its output is a set of logic levels which act as inputs to the control logic and the addressing circuitry of the microprogram store, and they must point to the start location of the microroutine which executes the function.
 The function of the flag or status register is not quite so obvious though its outputs also go mostly to the control logic. The so called 'conditional' instructions depend on the state of the flag register at the conclusion of the previous instruction as to which alternative action they take. For instance the instruction 'jump to address n if accumulator contents are 0' requires that if $C(Acc) = 0$ then the address n must be sent to the PC. If $C(Acc) \neq 0$ then the PC is merely incremented. The flag bit corresponding to zero accumulator provides a logic 1 if the jump is to be implemented but a 0 otherwise. This logic level is used in the logic networks to open some gates and close others to achieve the operation required.
 The choice of flag bits provided depends on the designer and his intentions as to how the system is to be used. Sometimes they are merely indicators, for the use of the program. Sometimes they act on the control logic, and sometimes they serve both purposes. Here again, it is stressed that the wise programmer reads the handbook very carefully before attempting to write a program.
 Our simple hypothetical CPU has had a few simple registers and one piece of complicated logic added but no extra highways. The provision of the extra logic unit which might well consist of a sophisticated incrementable address buffer should prove cost effective since it would be in use continuously and save a number of microsteps in every instruction. Pointer incrementation now poses few problems due to the

44   *Understanding Microprocessors*

the address buffer and is likely to be efficient. The separate incrementation logic no longer interferes with the flag register and the execution of conditional instructions.

The central registers would be cheap to provide and have many uses but they would enable several program processes to be run in parallel with a minimum of reference to the main data storage. The existence of the two auxiliary registers allows the data highway to be available for other activity while the ALU is operating, thus enabling some parallelism.

It seems, then, that the upgrading of our hypothetical CPU has achieved much of what is required without unreasonably infringing the design constraints imposed by the technology. How true this conclusion is must also be hypothesis since we cannot run the device and try it. What we can do, however, is to compare it with some real microprocessor CPUs that have proved themselves in use.

SOME CURRENT MICROPROCESSORS

The Intel 8008

This was the forerunner of the microprocessors described in this book. In fact it was the forerunner of all the 8-bit microprocessors and to Intel must go the credit of being the first in the field with a commercial 8-bit microprocessor. From the 8008 and the experience of using it, Intel and all the others in the business learned a lot: the second generation devices were very much better having profited from this experience.

At the time when the 8008 was being designed chip manufacturers were still trying to keep pin-outs down to 14 or 16 terminals, conforming to the standard DILIC encapsulation (dual in line integrated circuit). The 8008 did conform very nearly, having only 18 pins. Figure 4.2 illustrates the pin layout. The most important feature of this first generation microprocessor, emphasised by the diagram, is that there is only one 8-bit bidirectional data bus and no other data channel. All data transactions with the outside world, including those with the system

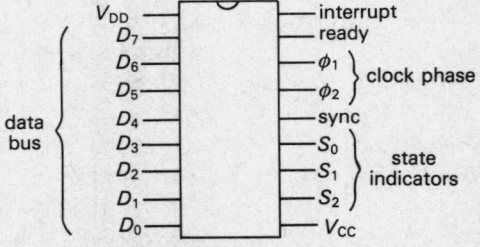

Pin allocation of Intel 8008: surely the least possible!

Figure 4.2

memory, have to be via this highway: it has to be time shared between transactions. The data routings between various sources and destinations inside and outside the chip have to be synchronised. The two-phase clock waveforms applied to terminals ϕ_1 and ϕ_2 control the speed of operation but there also need to be three 'state' terminals S_0, S_1, S_2; the outputs from these are coded to indicate to the external logic which one of six internal states is prevalent at any instant.

Figure 4.3 shows the register and highway interconnections of the 8008. A brief consideration of the processes involved in fetching and executing a simple instruction will demonstrate quite clearly why all the consequent second generation of microprocessors have an augmented pin-out to allow for a separate address bus.

The fetch instruction phase requires that the contents of the PC are sent to the memory address registers. Since the PC is 2 bytes in length, this transfer entails sending the high and low bytes individually, and in succession, to the memory address registers. The high byte is sent to the more significant and the low byte to the less significant register. The decoding of S_0, S_1, S_2 is used to control this. The store puts out the instruction byte. If it is a slow store relative to the processor clock rate, it puts out first a 'ready' signal which prevents the CPU from trying to read the data before it is ready. When 'ready' goes up, the CPU moves to its 'read' state, opening its data port and reading in the instruction byte to the control register through the internal highway.

If the instruction 'read in' demands an immediately addressed operand, the PC is incremented and exactly the same procedure takes place to read in the operand byte but, this time, the data is routed into the appropriate internal central register. This completes the fetch instruction and fetch operand phases. The execution, the actual processing activity, is rather simple compared with these. It is not hard to see why the designers provide seven central registers. Judicious use of these can save a lot of store transactions.

The next important feature to note is the large pointer stack, which provides the program counter and seven additional levels. With such a restrictive highway structure, the problem naturally arises of how to deal with subroutine jumps. Pointer addresses can be assembled in the two central registers marked H and L. An instruction sends them directly to the program counter and causes the existing PC contents to be pushed on to the stack. The return from the subroutine merely entails a pop or unstack instruction which brings back the original contents to the program counter. Having several levels of the pointer stack allows subroutines to be 'nested', that is, a further subroutine can be called during the execution of a current one, and so on. Naturally this process must be performed with circumspection—if too many pushes are performed data can be lost.

The 8008 also uses the pointer stack to deal with peripherals and

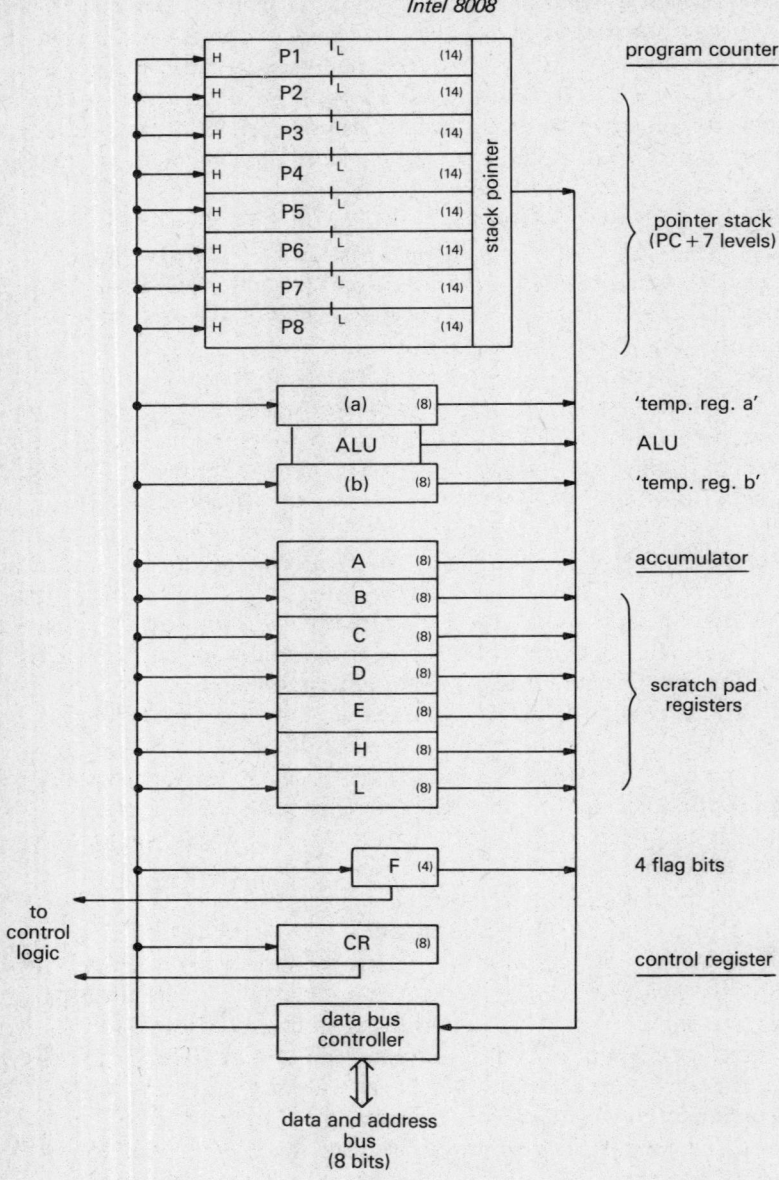

Figure 4.3

interrupts. Instead of each peripheral merely having an arbitrary address in a vacant area of the store address field it has, as an identity, an address in page 0 which is also the start address of the service routine for dealing with it. Thus, if the peripheral device signals an interrupt the

CPU merely completes its current instruction, stacks the current PC value and reads in from the data bus an identity byte placed on it by the calling peripheral (synchronised once more by the appropriate coding on S_0, S_1, S_2.) This identity byte becomes the contents of the PC register and the servicing routine commences. At the end of the operation the stack is popped and the processor resumes its operations from where it left off when interrupted.

The 8008 conforms painfully to the Wilkes microprogrammed processor concept: manifestly this concept is not very applicable to a device having a word length of only 8 bits. Nevertheless, the 8008 was the trail blazer; true, it was slow, but it provided a useful computing device which, when harnessed into a system, gave effective service at a cost an order of magnitude less than the mini computers of its time. A great many experimental systems were built round it and a number of applications were successfully developed: it proved reliable and effective. Incidentally it provided a useful apprenticeship for designers of future, easier, microprocessor systems.

It was followed by a range of second generation 8-bit microprocessors, perhaps the most notable of which was its own direct successor, the Intel 8080.

Intel 8080

The register and highway arrangement is shown in figure 4.4. The first difference, and design advance from its parent 8008, is that the 8080 has a separate 16-bit address output. This connects the 16-bit pointer registers through the address buffer to the external address bus. This allows the store to be addressed and accessed in a single cycle. The pointer registers are incremented by their own logic without affecting the ALU. The second pointer register is a stack pointer allowing a stack of any length to be set up by program anywhere in store.

The central register group is shown in figure 4.4 in the way it is described in the manufacturer's literature. At first sight it seems somewhat different from our hypothetical CPU arrangement, though it does, in fact, conform. There is one accumulator A; Registers B and C, D and E and H and L are considered as pairs. The pair H, L is little different from the registers of the pointer group and could well be associated with them. In fact each of the six central registers is available to the programmer as a general purpose or scratch pad register but they can also be used in pairs as 16-bit pointers and they can be dealt with in pairs by single instructions. This allows considerable flexibility since it allows as many as three index pointers to be used if the programmer so chooses, as well as the stack and PC pointers.

The peripheral interrupt arrangement is also a logical advance over the 8008. Instead of the PC stack, the interrupt arrangement is to allot

Intel 8080

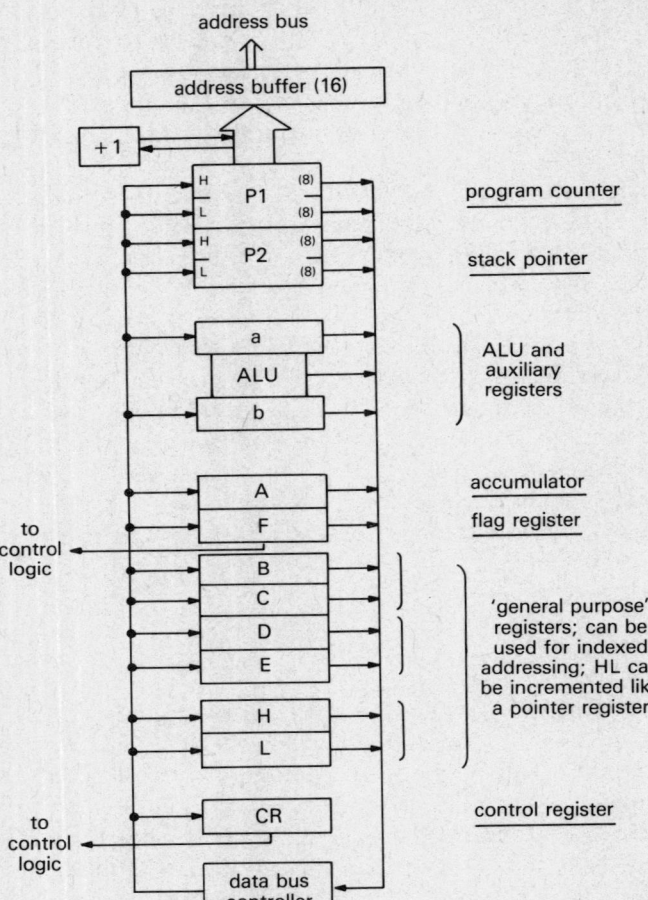

Figure 4.4

eight locations in an interrupt page. Thus when the interrupt is acknowledged all that is required is that the peripheral logic puts on to the data bus one of eight 3-bit codes, giving access to one of the prearranged locations, in which a jump instruction can be placed to any suitable location, for the service routine required. This considerably simplifies the peripheral logic interface.

Motorola 6800

Figure 4.5 illustrates the register and highway arrangement of the 6800 which has proved itself a well-liked and effective microprocessor. It is

Practical CPUs 49

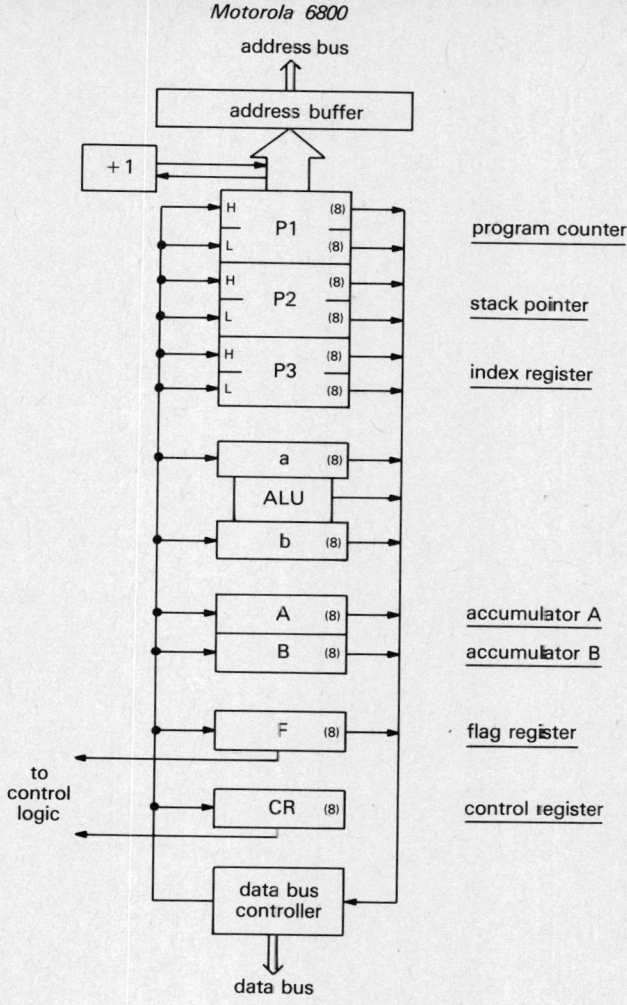

Figure 4.5

considerably simpler than the 8080, which might well account for some of its popularity. It conforms quite closely to the structure of the hypothetical CPU we have considered and also to that of the Digital Equipment Corporation 'DEC PDP8', the archetypal and most successful minicomputer. Although the structure is so simple the device has a powerful instruction set: both central registers A and B can be used as accumulators for most operations. The designers have placed less importance on additional central registers but have provided an index register in the pointer group and a wide range of memory reference instructions giving considerable addressing flexibility.

The 6800 has two interrupt lines; NMI, which stands for nonmaskable interrupt, is of particular use for emergency interrupts such as power failure. The normal maskable interrupt causes a jump to the contents of store address FFF8/9. NMI causes a jump to the contents of FFFC/D. There is also a restart which causes an automatic jump to the contents of FFFE/F, the highest possible store addresses. Both interrupts cause the automatic saving on stack of the contents of the working registers. Restart, naturally, performs the reverse process. Since the cause of interrupt is not identified, the interrupt service routine must first poll the status values of the peripherals to establish which one is calling before a jump can be made to the particular routine appropriate.

The National SC/MP

The register and highway arrangement is shown in figure 4.6. This is an elegantly simple microprocessor. It has a curtailed PC of 12-byte length only, but allows for the most significant 4 bits of a 16-bit address to be entered into a 4-bit page register which effectively forms the upper half of the H PC register.

The most significant difference of the SC/MP from most other 8-bit microprocessors is that the extension register, which acts as a second central register, can also be right shifted in serial mode. The most significant bit of the extension register is accessible from an input terminal and the least significant is connected to an output terminal. This allows serial input data such as from a teletype to be read in directly. A delay instruction is provided to facilitate this process. A delay of any number of machine cycles can be set up simply and called for use when reading in data bits to accommodate the speed of the input device. Likewise data for a teletype or similar device can be set up in the extension register and output serially at the required data rate. This facility allows SC/MP or 'Scamp' to be workable in the simplest possible configuration with no interface adaptors or similar provisions.

The Zilog Z80

This is the first truly third generation microprocessor. The register and highway connection closely resembles that of the 8080, which should not be surprising since it was designed by the same team. The most important structural differences are shown in figure 4.7 and are as follows.

(1) There is an additional array of central registers A', F', B', C', D', E', H', L'. This array with its contents can be simply switched into operation by a single instruction, replacing the array of central

Practical CPUs

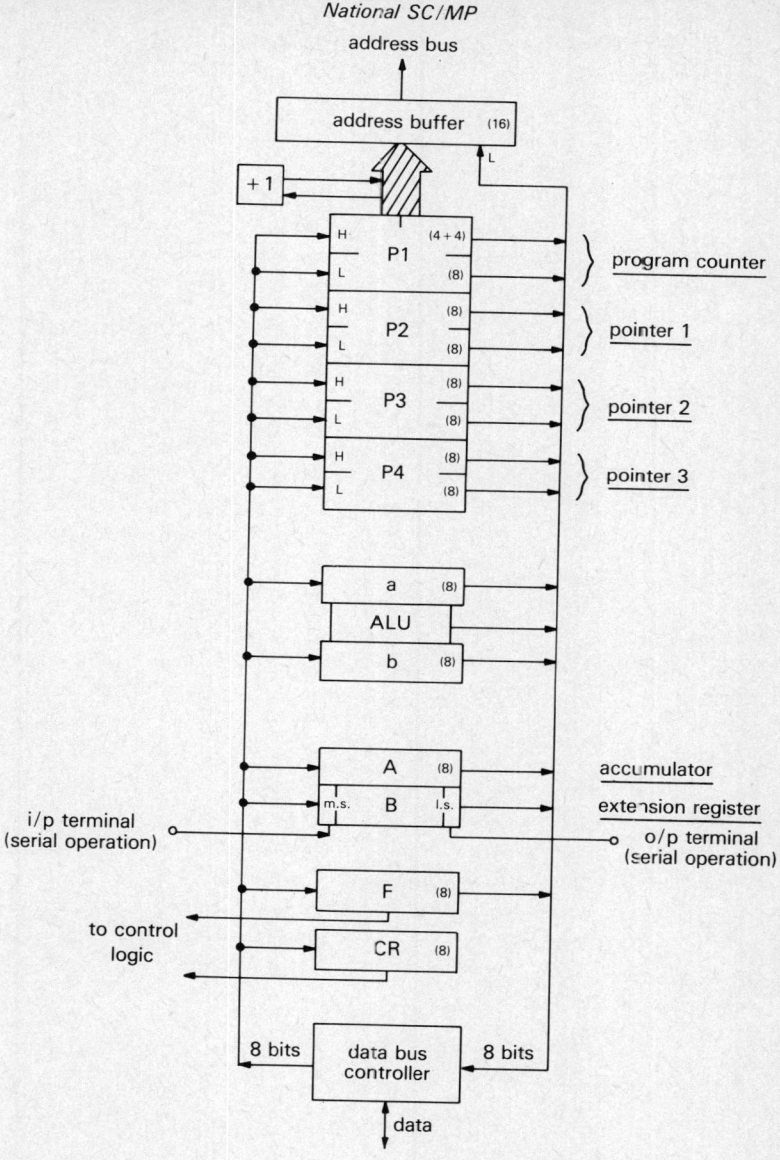

Figure 4.6

registers *A* to *L*, and can likewise be switched out of operation again. This provides a fast means of switching from one program to another.
(2) Two index address pointers IX and IY are provided.

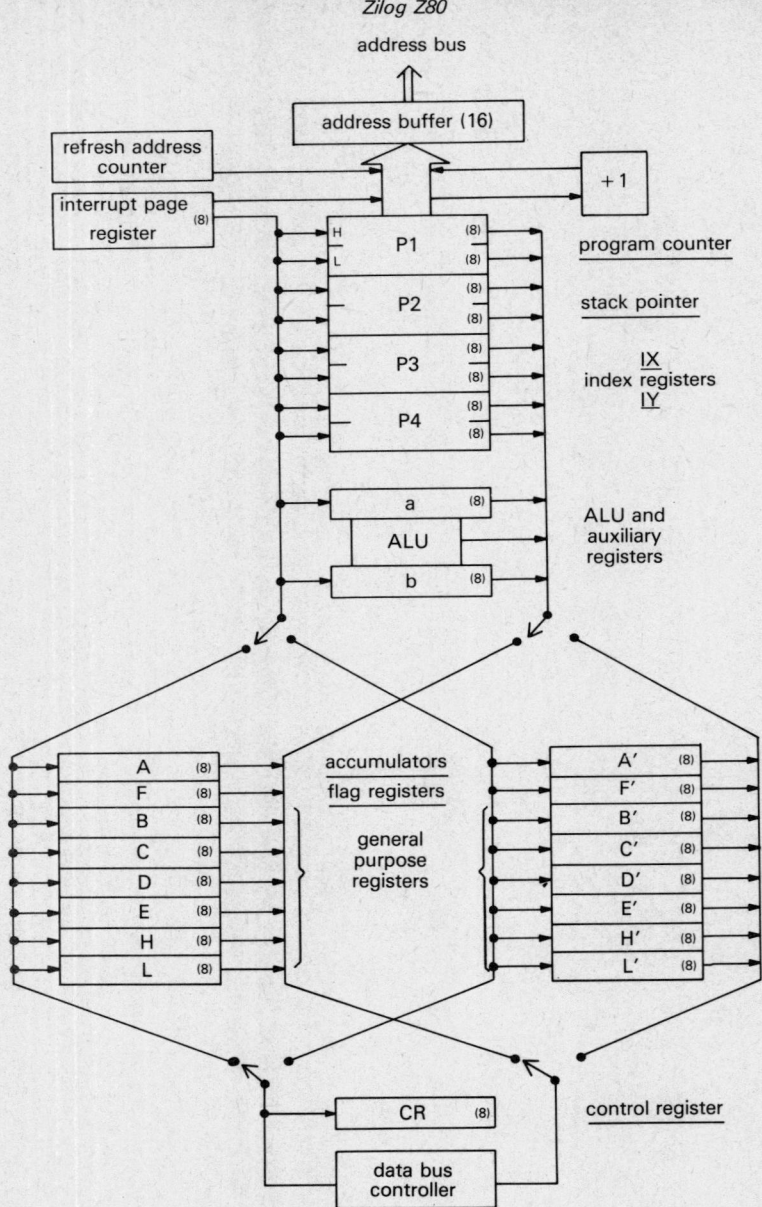

Figure 4.7

(3) An interrupt page register is provided. This 8-bit register is loaded with the page number allotted to the interrupt addresses.
(4) An 8-bit refresh address register is provided which automatically addresses dynamic store blocks if they are used, saving the system designer from having to provide refresh circuitry.

The Z80 is microprogrammed so that it can run any program written for the 8080 as well as do much else besides. It also has a set of specific input and output instructions.

5 Programming

All data processors, large or small, have to be programmed to make them do anything useful. In the early days of computers, the writing of programs began as contrivance and developed into art. The art became rationalised into a scientific technique as more computers appeared and more people gained experience. Today it forms the major part of computer science. Although some aspects of modern computer science are abstruse and academic it is, nevertheless, a very practical study. The actual job of programming always forms a major part of that study. A lot of programming at the lower levels is done—and must continue to be done—on the basis of ingenuity and common sense.

This chapter explains what programming is and how it is done. A single chapter of a book cannot teach anyone to program. To become a competent programmer diligent study and, above all, practice are required. It is a study that can last a lifetime.

Machine Code

Whatever methods and techniques are used for developing and writing them, programs must end up as strings of binary patterns that can operate the computer. These patterns of the instructions and data are rationalised to form a code and this is called *machine code*. This code is peculiar to any particular machine and depends on its structure and logic.

Consider a simple short list of instructions and data for the trivial task of adding together two numbers and noting the result. From our knowledge of processor systems derived from earlier chapters we can write

(1) Clear the accumulator
(2) Add the contents of location m to the accumulator
(3) Add the contents of location n to the accumulator
(4) Store the accumulator contents in location n

If we insert our two numbers, x and y, in locations m and n, and run the program we should end up with $x + y$ in location n.

Programming

Each instruction after the first is memory reference and takes the form

| function | operand address |

Suppose the program is to be run in a microprocessor of the type we have been considering: An 8-bit byte forms the instruction word and two further 8-bit bytes will be needed to provide the operand address. Generally the instruction byte is partitioned into fields: suppose the first 4 bits define the function, bits 5 and 6 the addressing mode and bits 7 and 8 the working register. The list of bytes to be stored as the program could well be

```
1 0 1 0 0 0 0 0   Clear accumulator
0 1 0 0 0 1 0 0   Add, direct, accumulator
0 0 0 0 0 1 1 0   High byte of address m (page 6)
0 0 0 0 1 0 1 0   Low byte of address m, (byte 10)
0 1 0 0 0 1 0 0   Add, direct, accumulator
0 0 0 0 0 1 1 1   High byte of address n (page 7)
0 0 1 0 1 0 1 1   Low byte of address n (byte 43)
0 0 1 0 0 1 0 0   Store, direct, acccumulator
0 0 0 0 0 1 1 1   High byte of address n
0 0 1 0 1 0 1 1   Low byte of address n
```

The same listing for a larger, 24-bit computer, assuming a similar function code, could be written

```
1 0 1 0 0 0 0 0 0 0 0 0 0 0 0 0 0 0 0 0 0 0 0 0
0 1 0 0 0 1 0 0 0 0 0 0 0 1 1 0 0 0 0 0 1 0 1 0
0 1 0 0 0 1 0 0 0 0 0 0 0 1 1 1 0 0 1 0 1 0 1 1
0 0 1 0 0 1 0 0 0 0 0 0 0 1 1 1 0 0 1 0 1 0 1 1
```

The generation of programs in this notation is painfully tedious and highly error prone. In the early days, however, full length programs for jobs like payroll or trigonometrical table calculation were written like this, often on machines with 32-bit word lengths or more.

It soon became obvious that this form of programming could be divided into two processes: The first was essentially a human activity, that of deciding how to do what had to be done: this was better performed without the constraints of binary coding. The second process was the translation of the human formulated program into binary and this was seen to be exactly the sort of task that computers could do best.

The human activity soon became formalised into the use of mnemonics for the functions and decimal numbers, or symbols which could be translated into binary numbers when the numeration had been worked out. The simple routine above could then be written, perhaps, as

```
CLA      /Clear accumulator
TAD m    /Add contents of m
TAD n    /Add contents of n
STO n    /Store in n
```

A special program first had to be written in binary code the hard way, which identified the mnemonics as functions and translated them into binary code; the symbols for numbers could be held until, at a later stage, they too could be supplied to the computer as decimal numbers for conversion to binary. The computer then produced the completely binary version by a process that came to be generally known as *assembly;* the program for achieving it was called an assembly program or merely an *assembler*.

Assemblers

Virtually all machine code programming is today written in assembly codes, sometimes called assembly language. Let us consider a simple example. Suppose functions are allotted mnemonics as follows

Function	*Mnemonic*
Clear accumulator	CLA
Add to accumulator	TAD
Subtract from accumulator	SUB
Store contents of accumulator	STO
Jump to given address	JMP
Jump if accumulator contents = 0	JMZ
Stop	HALT

The program is written line by line, each instruction taking one line. The line number is given in the left hand column. On the right, following the oblique stroke (slash) are written comments. These are disregarded by the assembler which merely stores them, as written, while the program is being prepared. The example program which follows is to sum fifty numbers stored as an array.

Line	*Function*	*Address*	*Comments*
1	CLA		/Clear accumulator
2	TAD	20	/Add current total
3	TAD	200	/Add next number from array
4	STO	20	/Store current total
5	CLA		
6	TAD	3	/Add Instruction 3 to accumulator
7	TAD	22	/Add one to address of instruction 3
8	STO	3	/Replace instruction in line 3
9	CLA		

10	TAD	21	/Add current COUNT value to accumulator
11	SUB	22	/Subtract 1 from COUNT
12	STO	21	/Replace COUNT value
13	JMZ	25	/Jump forward if accumulator = 0
14	JMP	1	/Jump back to start
.			
.			
.			
20		0	/TOTAL initially zero
21		50	/COUNT of numbers in array
22		1	/increment or decrement
.			
.			
.			
25	HALT		
.			
.			
.			
200	.....		/Array of numbers to be added
.			
.			
.			
249	.....		/

This program could be used to sum numbers in arrays of any length, merely by adjusting the COUNT value. Instructions 6 to 8 increment the address of the summing instruction through the array. Instructions 10 to 12 count the summations performed—the program loops. After 50, in this example, COUNT = 0 causing the JMZ instruction to jump the program out of its loop; the program stops the process with the total in location 20.

The program example assumes that there is only one central register and demonstrates how modification and process counting complicate the operation of a true single address computer. If the program were to be rewritten for a microprocessor having at least three central registers, then COUNT could be kept in one and the indexing of the instruction performed in another, leaving the main accumulator undisturbed to hold the total. The program would only be shorter by some STO and CLA instructions but the execution would be faster since this reduction would apply in every execution of the program loop.

The example can be used to point out several features of assemblers. Note first that the program was written starting at line 1 and all addresses were relative to this starting point. In actual operation it would not normally be loaded in location 1 onwards. A preliminary instruction would be written in some form like '*200' which would cause the assembler to add 200 to every address and thus hold the program in locations 200 to 448. Until the start address is fixed the program is 'relocatable'. This feature is used in sophisticated systems by a cunning program called the *operating system*. This program remains resident in the processor memory and performs managerial activities like store allocation, subroutine handling, calling the assembler and other service programs. It also maintains teletype contact with the operator, informing him of the state of the system and receiving his instructions in some convenient near-plain language code. Books have been written on the design of operating systems. A good operating system can make a data processing system efficient and easy to operate. It can also take up a great deal of storage space, hence microprocessor operating systems are usually simple, sometimes elegantly so, but limited in scope.

If we refer again to the assembler we can see that a characteristic of assemblers is that there is a one-to-one correspondence between the assembler program instructions and machine code instructions. The only exceptions are for instructions like that for the starting point. These are not translated and added to the assembled binary code but are executed immediately. These instructions are usually called *assembler directives*.

Assemblers can be simple translation programs or they can be sophisticated, giving a great deal of aid to the programmer. For instance, they may be able to cope with what are called 'labels'. In the example program, for instance, location 21 holds the COUNT value. In an assembler that is written to handle labels the instructions dealing with this count could be written 'TAD COUNT', 'STO COUNT'. This saves the programmer from having to look up the location of the count each time. A further facility that can be provided is to make the assembler handle simple expressions. In the example, locations 20 to 22 hold a set of values or constants concerned with running the program. Each could be labelled separately but it is often more convenient to treat them as a block and label them as such. Suppose we call location 20 'TOTAL'. We could then call 21 and 22 'TOTAL + 1' and 'TOTAL + 2', leaving the assembler to deal with the details.

Much of microprocessor programming is done in assembler code and most manufacturers provide assemblers. A major problem is that the actual assembler, if it is a powerful one, may be a long program and it is impractical to hold such a program in a small microprocessor. In this case the assembler is made to run on another larger processor and is called a *cross-assembler*. The program is then written in the microprocessor assembly code and assembled as an object program on the larger machine which holds the cross-assembler. The larger machine then generates the binary version on a suitable output medium such as paper or magnetic tape, together with listings of labels and data locations. This binary version can then be entered into the microprocessor to run it.

The assembler can do other useful things: it helps a great deal if it has diagnostic capabilities. Thus, if a program is entered for assembly that contains syntax errors the assembler can print out what they are. Syntax errors mean errors in writing the program so that the machine cannot execute it. Errors of concept, or in the algorithms for the process, cannot be detected by the assembler. Even if an assembler is not available the assembler code is a useful format for programming since it is based on the logic of the microprocessor and its register configuration and it provides a useful brief notation. It is probably for this reason that many people call assembly code machine code.

Ultimately a program must be tested by running it on the actual processor on which it is to be used. It is not likely that it will be correct in every detail the first time it is run. It must then be debugged, that is, the faults must be found and corrected and this often has to be done at the machine code level. The binary patterns in the various registers have to be examined and decoded. Since there is a one-to-one correspondence between machine and assembly instructions, this is not too difficult especially if the assembler listings are adequate and the program is well documented—problems arise if they are not.

Since the microprocessor is a single byte device and instructions may take 1, 2 or 3 bytes it is considerably more difficult to assemble or to debug and there is plenty of room for error in counting the number of bytes for looping or branching. One of the attractions of using an assembler program is that a good one can take care of much of this.

SUBROUTINES AND MACROS

Nearly every program has been written before and by somebody else. Naturally new applications arise and new programs must be written to suit particular conditions but even so they will consist largely of routines that have been used many times before. Even if we discount this, then there will still be routines like counts and similar devices that will be used repeatedly in any major program. To save useless repetitive work subroutines should be used.

A subroutine is a self contained segment of program written to perform a frequently used function. It is stored in the processor as an appendix to the program under execution. In large systems a library of such routines may be held and made available through the operating system. Typical well-used subroutines are those for performing division, generating logarithms or trigonometric functions or, in microprocessors, performing multiplication. Many peripheral devices are handled by subroutines as are processes of text editing and the conversion to and from teletype code or BCD. There are many subroutines of this kind used in an assembler program.

In order to use a subroutine the main program must be suspended.

The processor then 'calls' the subroutine, which means more precisely that it jumps to the start address of the subroutine and carries on processing from there. At the conclusion of the subroutine it 'returns', that is, it jumps back and carries on with the original program. This may seem simple enough but it does pose problems. In the first place the program count of the main program must be stored somewhere, since the program counter is to be used for running the subroutine. The data on which the subroutine is to operate must be placed so that it can do so. Data belonging to the main program that is not connected with the subroutine must be put somewhere where it will not be interferred with and it must be recovered again when the main program is restarted. Most microprocessors have call and return instructions which push on to the stack the current program count before starting the subroutine (jumping to it) and pop it afterwards. It is up to the programmer to determine what registers the subroutine will use and make provisions accordingly.

The subroutine idea seems, on the face of it, just a pair of jumps to bring in a piece of previously written program. What should be realised, though, is that it may well call a piece of well-tried and totally proved program. Even more important is the fact that the subroutine is only stored once but used many times. This distinguishes it from what is called a 'macro routine' or merely a 'macro'. This is a piece of code just like a subroutine but it is not called for use. When the program is assembled the routine is written in full in the main stream of the program, adding to the length of the program everytime it is used. Some assembler programs allow macros to be embodied, many do not. In fact macros usually get into microprocessor programs rather informally. The manufacturers programming manuals customarily give many useful examples of processes expressed in assembly code. The wise user embodies them wherever they are suitable, merely by writing them in. If, of course, they were to be used repeatedly it would be better to add them as subroutines and call them when required.

HIGH LEVEL LANGUAGE

As more and more use was found for computers it soon became obvious that most user programs consisted almost entirely of well-established subroutines and macros; the program writer selected them in the right order and provided the data. It was also obvious that assembly code programming was unnecessarily tedious and demanded knowledge of the processor and a good deal of skill. So-called high level languages were evolved to make programming acceptably easy for people who only wanted to make use of computers. The first two languages to appear that justify the term 'languages' were Fortran and Cobol: the former was for mathematical and scientific use, the latter for commercial. Each allows a problem to be expressed in a form fairly close to its normal form, in ordinary use. For example

$$X = \sqrt{\left(\frac{A^2 + Y}{2B}\right)}$$

is written as Fortran as

  X = SQRT((A**2 + Y)/(2*B))

and

'Compute net pay by subtracting previously calculated deductions' is written in Cobol as

  SUBTRACT DEDUCTIONS FROM GROSS-PAY GIVING NET-PAY.

These languages consist of a set of terms and rules for laying out and expressing processes. More recent and more powerful high-level languages are Algol, PL/I and Pascal.

The term high level in this context is ill-defined. A more expressive term is 'problem oriented'.

A great advantage of a high level language is that it is machine independent, that is, unlike assembler code, it does not depend on the particular computer on which it is to be used. The set of rules or specification of the language is published and, as far as possible, standardised. Each computer manufacturer then produces a program called a compiler, which can read in programs written in the high-level language and compile them, ultimately into binary machine code suitable for the particular processor. The compilation process is to dissect the object program, translate it into macros and subroutines and to allocate storage for its data. It is a highly complex process and demands the work of programmers and computer scientists of some skill if it is to be efficient. Like a good assembler, but more so, a good compiler must do a lot to assist the programmer, not only by relieving him of a lot of tedious work but also by providing diagnostic information where mistakes occur. Like an assembler, but still more so, a compiler takes up a lot of storage space in a computer.

The compiler takes in a whole program and processes it until it has finished, that is, until the whole program is in binary code, ready to run. This is fast and efficient where jobs are to be run in batches, or if only the final results are of interest to the program writer. However, many tasks that computers can do can be done better interactively; that is, the computer from time to time refers the problem back to the programmer, often in some special format such as a graph or table of data. The programmer then makes some decision and starts the next stage of the operation accordingly. It is difficult to compile efficiently for this kind of work hence another technique has been developed for dealing with high level languages which is called 'interpretation'. This is performed by a program called an interpreter. Like the compiler this is first run into the processor. When the object program is entered, however, the interpreter does not wait for it all but decodes it and obeys it, step by step. The process is normally about ten times slower than compilation, but it adds another dimension to computing. A special language called Basic has been evolved particularly for interactive use and many microprocessors are supplied with Basic interpreters. Basic is a simpler language although it is not unlike Fortran. It is easy to learn and a good interpreter program helps the user in the learning process. Naturally, if the interpreter is made to take up a minimum of storage, as it usually is for microprocessor use, then its performance must be greatly curtailed.

In Basic, the example program of the summation of the array of fifty numbers would be written

Line	Instruction
10	S = 0
20	FOR I = 1 to 50
30	READ A
40	S = S + A
50	NEXT I
60	DATA 1012 51615 321 ...
.	.
.	.
.	.
70	END

The program as written is practically self explanatory. Where explanation is needed in a program, a line is started with REM, short for 'remarks'. Thereafter the rest of the line is merely read in and printed out as written. Usually at the start of a program or a program segment, a title and explanation are written following REM.

The lines, or instructions, are numbered: Usually in interactive operation each line is given an advance of ten lines on the previous one. This allows the insertion later of intermediate lines if required. A very convenient feature of Basic is that a line number can be written at any stage but the interpreter will execute it in numerical sequence. To correct a wrong line, its line number is typed followed by the correct version: the interpreter then overwrites the old line with the new. Provided that a rather simple set of rules are rigidly observed, Basic is easy to write and execute. Given the set of instructions, most people can begin useful programming exercises after a few minutes and useful programming after not much longer. For this reason most microprocessor systems marketed for education, simple program development and for hobbyists, provide a Basic interpreter. This may be provided as a 'tiny Basic' of, say, 2000 bytes in a ROM, or as a highly sophisticated interpreter of say 14 000 bytes on a tape cassette, or some intermediate sized version. It is to be confidently expected that full sophisticated versions of Basic interpreters will soon be available in ROM at economical prices.

THE DEVELOPMENT OF MICROPROCESSOR PROGRAMS

In normal applications most microprocessor systems consist of the microprocessor itself, a program written in ROM and a limited amount of RAM, as work space and for holding current temporary data. A peripheral interface is then required to connect the microprocessor directly to whatever task it is to perform. Naturally, the program written or blown into the ROM must be correct in every detail and capable of dealing with every foreseeable circumstance of the application. It must work as soon as it is switched on. The program development for such a system is a somewhat complicated and exacting task and may be done in a number of ways, depending on the nature of the processor, the task and the facilities available for the programmer. However it is done, at some stage use is normally made of a device or system called an emulator: this may be improvised by the designer/programmer in the laboratory or may be a sophisticated device supplied as a development aid.

To emulate something is to do everything that the something can do and something more. A microprocessor system emulator must do everything that the microprocessor system is meant to do; in addition it must allow its operator to intervene in the process, by observing its operation, checking and correcting the program and inserting the corrections.

A complete emulator is usually a microcomputer based on the microprocessor to be emulated. To aid in the initial program writing, it will have an operating system written in ROM and may well have an assembler or even a Basic interpreter. The actual program under development will be written and held in RAM.

The emulator must have a conventional data input and output arrangement so that the program can be written, entered, tested and monitored. There will need to be a teletype or a video display unit (VDU) and keyboard at least; there may also be a cassette interface or a paper tape reader—in fact it needs the normal facilities of a general purpose minicomputer.

A minicomputer is normally equipped with a display console to facilitate running and debugging programs. This console displays, by sets of lamps, the current contents of registers: the PC, the accumulator(s), the memory buffer, and the machine status. It also has a set of switch keys as a pattern generator which can be used to set up binary patterns. These patterns can be deposited into the working registers. Further switches and buttons are provided for operating the system in single instructions or single steps within instructions. Sometimes provision is made for trapping, that is, setting a pattern or value on the switches for the PC, or perhaps a central register: this causes the system to halt when the pattern is detected. From our studies of the nature of the microprocessor we can see that such facilities just do not exist as an integral part of it. The emulator has to be made to cope with all these requirements. For instance, since none of the microprocessor registers is available to the outside world, the emulator operating system displays the contents of a register by taking over the microprocessor and causing it to output the contents required as an output operation. Likewise, for single instruction or single step operation the emulator contrives to halt the microprocessor usually by manipulating the clock waveform. There is much cunning and contrivance contained in the design of an emulator.

Once the program has been fully developed the emulator is arranged so that the process runs using RAM store, virtually in isolation from the emulator. The contents of the RAM store can then be extracted to provide the contents of the required ROM. This is blown and inserted in the system for final test. If everything has been done correctly, then the ROM and microprocessor should run as a separate entity.

If a system under design requires only a short program and a small ROM to control it, then it may well be programmed in assembly language and entered into the emulator in machine code. This is the method by which most hobbyists and engineers do simple designs and learn their trade. The emulator itself may be a rather simple system. It is a painstaking process and never a quick one.

For commercial and larger scale applications more sophisticated methods are used. Sometimes the initial program development is done using a simulator. This is a program which is run on a large computer. It is written in such a way that it can take in the machine code for the operation of the microprocessor and execute it exactly as the microprocessor would do it, though probably much more slowly. Such a program usually contains diagnostic and similar routines so that the program is not only tested for correctness of concept but also matters like process timing can be analysed. A simulator program is normally several hundreds of thousands of words long: it can only be run on a large computer and costs a great deal of money to develop. Some microprocessor manufacturers provide customers with the use of their simulator on their in-house large computer or through a data processing bureau, on a time-shared basis, over the data network.

To make the initial programming easier, cross-compilers are

sometimes available, similar to cross-assemblers but operable in a high level language. These are also run on a large computer.

The correction and rewriting of the program are facilitated by programs run on mini, or larger computers, called *editors*. These allow correction of the text by means of a display and keyboard: they put out on paper, magnetic tape, or disc, the edited version in correct format.

Although great skill and effort are expended on simulators and similar programming aids it is hard indeed to be perfect. It is particularly hard to foresee every eventuality. Furthermore, in real applications there are very often problems of timing and synchronisation that can only be effectively dealt with when the microprocessor is actually on-line to its task. So, ultimately, at the current state of the art, every application must finally be tested and run at machine code level. At this level sometimes even sophisticated emulators give distorted or erroneous results. There is no valid substitute for running a system under real conditions: the final correction and tuning of a complicated system requires people who are skilled in hardware, software and electronics. The technological microprocessor 'explosion' has occurred and is still occurring. Its impact is likely to be much more gradual than many people forecast. Rather few very clever people are involved in the technological development. Very many people indeed are required for the development of applications and most of these have yet to be found and trained. The application of microprocessors offers a fierce challenge and huge opportunities. Perhaps the readers of this book will make their contribution.

6 Interfacing to Peripherals

This imposing title simply means the connection of the microprocessor to the outside world. Up to now we have considered the microprocessor as a self-contained entity and paid little regard to what has to be done so that it can do something useful. The interface is the term for the hardware and software in the area between the processor and the 'something useful'.

Conventional computers are general purpose machines: they cost a lot of money and to justify this cost they are expected to perform a wide variety of tasks, often in quick succession and sometimes simultaneously. Their store capacity is necessarily limited. Switching from one task to another often means taking in long programs and great quantities of data. In many roles they also generate much data as output. Thus input and output handling is a major preoccupation of computer system designers. Quite often the efficiency of the computer is limited by its input and output capabilities and the main processor spends a high proportion of its time in the trivial activities of handling data rather than processing it.

There is a virtually obligatory set of input and output devices because of this data handling requirement, such as teletypes, readers, punches and printers. These devices, too, must be general purpose in that they have to convert data symbols, comprehensible to a wide population of users, into a symbology comprehensible to a wide range of computers. The only effective solution to this problem is to standardise the data forms and handling devices and this is done as far as is possible.

Many minis and some larger computers are also used as on-line process controllers of machines and automatic systems. These computers require not only the standard range of symbol conversion peripherals but also devices which convert from the internal symbology of the computer to what the system under control can provide and what it needs to control it. Like the symbol converters these, too, have become pretty well standardised. Typical of these devices are analog-to-digital and digital-to-analog converters. These convert between data in binary number representation and voltage or shaft rotation or some other physical parameter scaled to be proportional to the binary value.

The matter of data input and output provides one of the most

Interfacing to Peripherals

crucial differences between microprocessors and their larger brethren. The microprocessor is essentially a mass-produced and very inexpensive item. In the vast majority of its applications its users would not wish to program it, even if they were able to do so. The program for the application will be written indelibly in ROM. Should it need alteration the ROM will be exchanged for a new one. Normally the microprocesor embodied in an application will contribute so little to the total cost that when the application is no longer required the microprocessor will not even be salvaged. The input and output arrangements for a dedicated role of this kind will need to be capable only of dealing with the specific application. For instance a microprocessor used as the controller of a domestic washing machine will receive input from a few switches and buttons and probably some level and temperature sensors. Its output will be some voltage levels to control heaters, water cocks, motors and the like. It is a whimsical thought that it might accept washing instructions typed in from a keyboard and that it might print out a list of the laundry: most users will be content to supply the dirty washing and the detergent, press a few buttons and, 'hey presto!', take the clean washing out of the machine.

The microprocessor, though, because it is cheap and versatile, will also be used in roles where it does have to accept and output conventional data. One such application is in the peripheral devices of larger computers where it will perform data conversion and control the mechanism of the devices, thus relieving the larger computer of a great deal of trivial activity. Another dedicated role many microprocessors already have is to form the central processor of general purpose microcomputers. These are now being used widely as cheap minicomputers and, for many tasks, typically in schools, they are proving to be more than adequate.

Even though computers usually have special instructions and dedicated channels for input and output, the interfacing or attachment of peripheral devices is far from easy. The difficulty arises from the enormous discrepancy in speed between the electronics of the computer logic and the mechanisms that can deal with human capabilities of reading and writing data. The output of a character symbol from a computer to a printer entails the translation of a bit pattern, which may only exist for a fraction of a microsecond, into a mechanical movement that causes a character to be printed. Electronic activities are limited by the speed of light but mechanical ones are limited by Newtonian dynamics.

The nature of the microprocessor adds a further complication to these problems. It has no dedicated input and output channels; its only communication with the outside world must be arranged over its single input and output data bus, sharing it with its memory.

INTERFACING DEVICES

The designers of microprocessors realised from the outset that the great advantages of the single-chip CPU would soon be lost if it had to be surrounded by a mass of conventional logic similar to that which normally formed a computer peripheral channel. For instance, the original Intel 8008, once harnessed to its peripherals became quite hard to locate among the many chips that were required in the interface logic. Microelectronics designers therefore applied themselves to the task of designing special LSI chips to do the interfacing task. Their primary objectives were to design devices that would be easy to use and that would be sufficiently general to cope with the majority of applications. Meeting the latter requirement entailed designing chips that were not significantly less complicated than the CPUs themselves. It was not so easy to meet the requirement of ease of use. Certainly the special interface devices often eliminate the need for a good deal of logic design. However they still require careful consideration by the system designer and thoughtful treatment by the engineer who builds the actual system.

The Parallel Interface Unit (PIU)

Most microprocessor manufacturers have produced PIUs or PIAs (parallel interface adapters) which have quite similar characteristics. Some manufacturers made simplicity a prime objective while others went for sophistication. Some PIUs can be used with virtually any microprocessor of any make, while some can only work efficiently with microprocessors of their own family. The most common configuration conforms fairly closely to that shown in figure 6.1. It consists effectively of six registers forming two channels, each 1 byte or 8 bits wide. This is

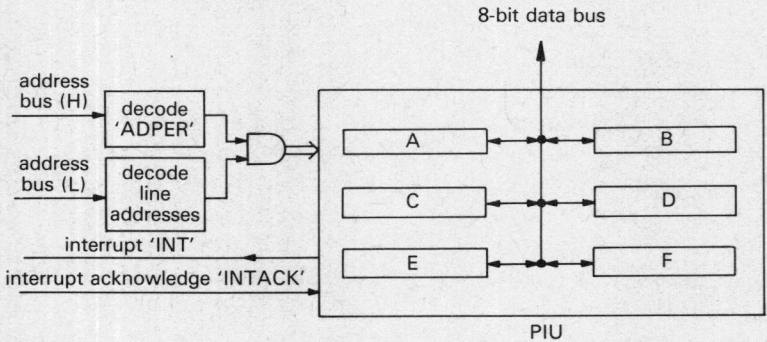

Figure 6.1

convenient since the microprocessor data bus is also 8 bits wide and most peripheral devices are designed to deal in 8-bit bytes.

In the figure A and B are the data registers for the two channels. C and D are the data direction registers while E and F are control registers. The device needs to be versatile since it is a chip and cannot easily be attached to switches even if it were desirable to do so. Instead it is programmed: the initialisation section of the application program contains instructions which are sent to the PIU in order to set it up for its particular task.

The first byte sent to the PIU is addressed to one of the control registers. The microprocessor address bus is normally 16 bits wide. It is convenient to allocate a page of addresses to peripherals: hence the high byte of the address is decoded once and for all by logic that generates a waveform or voltage level when it receives and recognises its page address code. This waveform is usually given an abbreviated name like 'ADPER' meaning address of peripheral. This being set enables the decoding of the low address byte which addresses a particular peripheral or a part of a peripheral channel.

We should note at this point how the microprocessor manages to do without special input and output instructions. Each peripheral channel or controller has an address as though it is a store location and it is seen by the CPU as a register, just as though it is a store location. Thus the CPU can perform an input merely by reading a byte of data from an address, or an output by storing a byte in an address.

Reverting to the PIU: the pattern loaded into the control register sets up the mode of operation of the channel. This register may also contain status bits which are set up by the PIU itself for the information of the CPU and can be read by it. The next byte of data from the CPU will normally be the set of channel direction bits which are set into the channel direction register. This register may be directly addressed by the CPU or through the agency of the control register. Each bit set into the register selects the direction of data flow into and out of the data register. Thus, an all 1s pattern may cause the data register to receive data from the CPU and send it out to the peripheral. An all 0s pattern would have the reverse effect. It is sometimes convenient to arrange for the data register to be outward on some lines and inward on others. Whatever arrangement is required can be set by the bit pattern. Often, once set up the channel remains unchanged for the whole operation thereafter, but in some applications like reading from a keyboard the channel operation is changed continually by program. Typically a bit set into the control register may reverse all the direction bits in a single operation. Resetting the bit later restores the *status quo*.

The actual process of transferring data depends on the arrangement of the microprocessor system. Suppose the CPU requires to output a byte of data to a peripheral attached to the channel ACE in figure 6.1. It must first determine the status of the peripheral, that is, whether it is ready to receive data, or even whether it is switched on! To do this it will need the contents of the control register E. It will test the appropriate bit and if it indicates that it is ready, the CPU will do a store operation, of the byte to be output to the data register A, already presumably preset for output operations. Once the PIU receives this byte of data it will put up a 'busy' bit in its control register E until it has disposed of the byte of data to the peripheral device connected to the channel. The presence of the busy bit should prevent the CPU from trying to send more data until the peripheral is ready to receive it.

An output of data is at the behest of the CPU and provided the wanted channel is not busy it can dispose of the data. If not it must wait; this it may do by running a very short program loop continually testing the status bit and jumping back. If it is merely trying to output data as quickly as it can then this is satisfactory. Otherwise it is wasteful of computer time. In this case the program may carry on with some other activity and try again later.

Data being input to the microprocessor system offers a different problem because, in this case, the transfer requirement occurs at some arbitrary time. To deal with this problem an interrupt strategy is used.

When data is ready for input to the processor, the PIU puts a level (usually a pull down to zero) on the interrupt line, commonly shared with all other peripherals. The CPU tests the interrupt line at the end of each operation. If it detects that there is an interrupt, it suspends its current operations and goes into an interrupt servicing routine (ISR).

The interrupt servicing routine must first determine which peripheral is calling for attention. It therefore 'polls round', that is, reads in turn the contents of each channel control register until it finds one with an interrupt bit set. It then branches to the particular servicing routine for dealing with the identified device and acknowledges the interrupt by putting a logic level on the INTACK line. The presence of this level enables the PIU to open the gates and leading from its data register on to the data bus. The CPU can then read the byte of data from the register and take off INTACK, informing the PIU that the transfer is complete.

The basic operating principle of these data transfers through the peripheral interface is called a 'handshake'. In each case the device initiating the operation must first signify its intention, and test to see if it is possible. When it receives a signal to say that the transfer can take place it causes the transfer and stays in the transfer state until it receives notification that the transferred data is accepted. It is the careful arrangement of this handshake principle that allows devices with widely different speeds to intercommunicate. It is used extensively in all computer systems and is of particular importance for microprocessors since their single data bus is shared by so many devices, including the memory.

The transfer mechanism described serves for dealing with data in 8-bit parallel bytes. Not all the 8 bits need to be used. If just a single voltage signal is required then the same device can still be used but 7 of the bits do nothing. A particular peripheral device may have as many as eight single on/off controls. Each can be allotted a bit position and the same system can put out these bits as ordained by the program one, or several, at a time. Thus the standardisation of the interface round 8-bit data transfer devices is not a limitation on flexibility.

Sometimes in on-line control applications the input and output required are analog rather than digital: for instance, a microprocessor may need to read in data from temperature sensors and put out voltages proportional to the amount of correction or control required. In this case the PIU is connected to the analog/digital converters. The microprocessor can then output voltages or sample them, working to a precision of 8 bits, that is, one part in 256 or about 0.4 per cent. The microprocessor may be able to do the timing of these operations by program but this is usually difficult since its internal operations may vary from instant to instant. The strategy, then, is to use what is called a 'real-time clock'. All this need consist of is an oscillator which puts out a square wave form at a suitable frequency. This waveform may be applied to the converters which can then call the processor when they are ready, or the processor itself may test the clock to determine its own timing. Sometimes the clock oscillator is attached to the interrupt line like any other peripheral.

The Serial Interface Unit (SIU) and UAR/T

Teletypes and video display units do not transfer bytes of data in parallel because they are communications equipment and operate over one pair of wires in their communications mode, for instance, over telegraph circuits. Since teletypes and VDUs are virtually indispensable as data handling units in program preparation and in many control and programmed operations, most device manufacturers market serial interface units or adaptors.

Before the advent of microprocessors, microelectronics manufacturers found a particularly suitable device for LSI namely the universal asynchronous receiver/transmitter (UAR/T). Until then, every computer, large or small, had to have a teletype attached to it and for each of them a logic network was designed for interfacing them to each other. The LSI manufacturers saw that there was scope for a standardised LSI device: the teletypes were standard, even if the

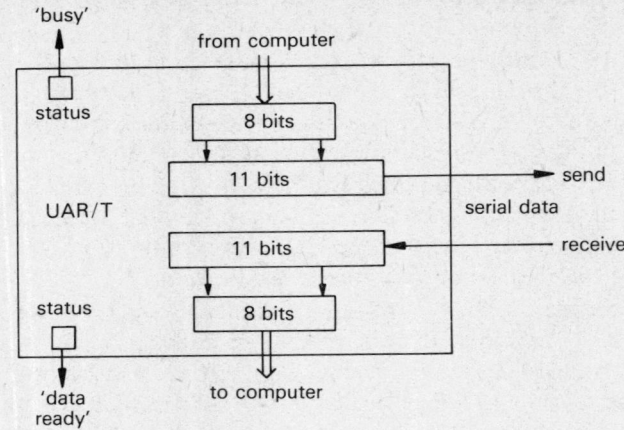

Figure 6.2

computers were not. The basic structure of a UAR/T is shown in Figure 6.2. It is divided functionally into halves, a send half and a receive half. Although it is possible to operate teletypes in duplex mode over two lines it is easier to interface them to computers a short distance away using two separate circuits and four or six wires.

Consider the output of a teletype character in ASCII code. The character is first assembled as an 8-bit byte in a central register. The CPU checks the status of the UAR/T by testing its status flag. If the device is not busy it transfers the ASCII byte into the UAR/T 8-bit data register as a parallel transfer. This sets the busy flag and causes the UAR/T to start operation. First the 8-bit code character is shifted in parallel into the 11-bit send shifting register. This adds start and stop bits to the 8-bit code pattern making it into an 11-bit byte, which is what a teletype deals in. The contents of the 11-bit shifting register are then right shifted out of the 'send' register to the output line terminal at teletype speed, that is, 110 baud or 110 pulses per second. It thus takes one tenth of a second to transmit one character. The signal pattern output is indistinguishable from that put out by a teletype.

The receiving process is merely the reverse of this. When the receive line terminal senses the presence of a start pulse it starts its timing mechanism—a series of counters driven by an oscillator. Midway through the pulse time it samples it to ensure that it is still present; at the end of one pulse time it left shifts the receive register, reading in the start pulse. Each succeeding pulse level is then sampled at mid pulse time and shifted in. When the start pulse arrives in the eleventh register stage the UAR/T has collected a whole character. The UAR/T checks the presence of the start and stop pulses and may also check parity: if it finds no error conditions it transfers the 8 ASCII code bits into the output data register and puts up a 'data ready' flag. This can be used to interrupt the computer which must then empty the data register within the 'crisis time' of one tenth of a second. This time allows even a microprocessor to do 20 to 50 000 operations.

The UAR/T is called asynchronous because it uses the start/stop principle that teletypes use. These machines, spread across continents, cannot have a common clock to control their speed. When a key is pressed on a teletype the device starts to operate at that instant and outputs its data at 110 baud more or less, depending on its motor speed, age and lubrication. The receiving device detects the start pulse and samples the incoming pulse train at mid-pulse times regulated by the speed of its own motor. Provided that the machine speeds do not differ by more than about 4 per cent data is successfully transferred.

To make the UAR/T universal also, it is possible to select a variety of modes of operation by putting appropriate voltages on control terminals. By appropriate selection the device may work with 5, 6, 7 or 8 bit codes, with even, odd or no parity generation or checking. By supplying its basic clock oscillator externally its speed can be controlled; data transfers between computers can be arranged at speeds of many thousands of characters per second.

The SIU is a development from the mix of a PIU and a UAR/T. Like the PIU it is under program control and made to be as flexible as possible.

By the provision of the PIUs and SIUs the microprocessor designers have removed much of the necessity for logic networks and logic design in the processor interface. There are other units which go with these: real-time clocks and interrupt priority decoders are typical. However there is a limit to what standardised devices can achieve. They must be standardised because they must be mass produced and have a mass market. To be standardised and yet to cover a really wide range of applications they need to be complicated. The more complicated they get the more difficult they are to use and the more room there is for error. The very clever microcircuit interface devices have taken a great load off the interface logic designer but transferred much of it on to the interface system designer and programmer. Outside the peripheral interface devices must still come the electronic interface to meet the requirements of logic level changing, power switching and amplification demanded by the peripherals whatever they may be.

7 Microprocessors in Perspective

The fundamental principles underlying the functioning of all digital computers are the same whether they are micros, minis or mainframes. Much of the understanding which this book is intended to impart is also applicable to all sizes of computers.

The differences that exist between the various genera of computers arise largely from their different roles or applications and the methods of their construction. Two factors mainly single out the microprocessor from its larger brethren. The first is its cost, which is so small that a whole host of applications for it become practicable which have not been so for the others. The second is its fabrication which concentrates it on very few planar silicon chips thereby restricting the provision of internal highways and interconnections with the outside world.

The type of microprocessor considered throughout the earlier chapters of this book has been the conventional 8-bit device which now dominates the field both in interest and in the number and range of its applications. In this chapter we shall look briefly at other forms and their peculiarities and possibilities.

THE SINGLE-CHIP MICROCOMPUTER

The processor systems we have considered up to now have consisted of a single chip containing an entire CPU but with the memory and peripheral interface devices on other chips connected to it. As the technology of getting more and more devices on a single chip has developed it has become possible to get a fair-sized memory as well as a CPU on to a single chip. Naturally in the early stages of such development both the memory and CPU have been of very restricted size and performance. Nevertheless a host of applications have been found which only require this: many single-chip systems have been satisfactorily developed and exploited. Made in mass-production quantities these devices have become very cheap indeed for what they can do. A typical example is the chip that forms the basis of the cheapest pocket calculators. This single device provides not only the calculating facility but also detects and decodes the keyboard input and drives the illuminated output display. It must therefore contain not only the CPU but also the ROM and RAM memories and the interface units.

Since these single chips made in large enough quantities can be priced in tens of pence rather than pounds they can be embodied in products that themselves cost very little. We can envisage, therefore, these devices greatly outnumbering any other kind of microprocessor very soon even if they do not do so already.

Currently, single-chip microcomputers are found largely in pocket calculators, as mentioned before, but also in the multitude of TV games and the like which have become commonplace. They are beginning to appear in instruments like multi-range test meters and they will doubtless be used in many small automatic devices for use in both industry and the home; an example of the latter use is in the control of microwave ovens—many more uses must follow.

For most people, other than microprocessor enthusiasts, these single chip devices will be of little interest in themselves. This is because, by their very nature, they are intractable little things. To be cheap they must be manufactured in very large quantities for totally dedicated roles. Their programs must be written, or masked, into their ROMs during manufacture. These programs must be developed and tested in every detail before manufacture and this is necessarily a complex and painstaking task and not anything to be taken lightly. Probably all this work will in future be undertaken by a few specialist consultants and the manufacturers themselves.

The normal process of design and manufacture must proceed somewhat on the following lines. First of all a manufacturer of some automatic device must establish the advantages of including a single-chip microcomputer in his product and that he has a large enough market to outweigh the heavy capital cost of the system development. The exact task of the microcomputer must then be totally and unambiguously specified; this may well be a sizeable and exacting task in itself.

Once the specification has been firmly established, the program that will ensure its operation must be written and somehow tested: this must happen before the actual microprocessor exists. It is normally achieved by using a large computer. this computer is provided with two specially written programs—a simulator and a cross-assembler or cross-compiler.

The simulator program is written so that it interprets a program for the microprocessor exactly as the microprocessor would perform it. It has to test the program for logical correctness and, usually, timing. It must interpret every idiosyncracy of the microprocessor. It has to be designed and written with great skill. It may be voluminous: two hundred thousand bytes is a reasonable size for such a program. Naturally it costs a great deal of time and money to develop and test.

The cross-assembler or cross-compiler is usually simpler and more conventional. This converts the program written in assembler or in high level language into the binary machine code of the microcomputer. Typically it may also require to be about 100 k bytes in size. It is not feasible for many people to produce or possess such programs. They are only reasonable for a specialist consultant enterprise or for the microcomputer manufacturer who can then use them for a large number of applications to affect their capital cost.

These two programs produce the binary code for the microcomputer ROM. A further stage of testing is carried out using a system emulator before final production is started. This is a special version of the microcomputer chip containing essentially the identical CPU, but analogous to a more conventional microprocessor. This is arranged to operate with an external ROM and a standard output arrangement. The program is run using an external RAM or ROM and the performance of the device carefully checked. When the design team are certain that their system truly works, the ROM and actual output details are masked into the production chips. The first batch must then be tested thoroughly; if all goes well, the production of the whole order can then proceed—if not there are tears and wringing of hands and a good deal of wasted work—and money.

There is a cheaper way of designing these devices but it requires skill and patience. For very simple systems it may well be satisfactory to start at the emulator stage. Then a test RAM or EPROM may be used that has been written with another device, possibly a microprocessor dedicated to ROM making. After satisfactory tests the ROM program may then be passed to the manufacturer for production. This method does without the heavy outlay on special software but is very time consuming except for truly simple systems. It would hardly be applicable for designing a sophisticated calculator but might well be economical for control of a small tool or meter or domestic device. In educational and similar establishments it provides a useful training exercise for those who may end up in a specialist design team, or who may depend on the use of single chip devices.

More recently, with the advance of EPROM technology, a number of single-chip microcomputers have been produced which have an EPROM store and can be programmed individually. This type of device will necessarily cost much more than the mass produced ones but it does allow single chip computers to be used in small quantity manufacture. These should find considerable application in specialised measuring instruments and the like. Since they have erasable stores that can be rewritten they can be used for successive dedicated tasks. They could well find extensive use as device controllers in laboratories and institutions which have the necessary design and EPROM blowing facilities ready to hand. Since their cost new is quite small it seems unlikely that they will frequently be salvaged when their original role becomes obsolete.

MICROPROCESSORS OF 12 AND 16-BIT CAPABILITY

The 8-bit microprocessor is inconvenient when it comes to dealing with precise calculation such as in continuous process control. Today there are an increasing number of 16-bit devices becoming available and for some time there has been a 12-bit CPU, the Intersil IM6100. This device is program compatible with the famous and widely used DEC PDP8 and as such it can use the comprehensive software written for that machine. Its designers intended it to be used on systems developed on the PDP8 but which could not justify the cost of a dedicated PDP8 to run them. It has the particular advantage that a great number of people are familiar with PDP8 and can use the micro version with little difficulty.

The reader might well wonder why the IM6100 does not threaten the PDP8; it is simply that the PDP8 is a fully fledged minicomputer with all the normal facilities of such a machine. It is also considerably faster. If one is tempted to emulate the PDP8 as a minicomputer system using the chip in place of the PDP8 CPU, the total cost of the peripheral and display attachments very soon bring the project up to a cost comparable with the mini, achieving considerably less performance at not much less cost. This example illustrates very well the essential difference between minicomputers and microprocessors. The micro is cost effective when applied to a dedicated task with no trimmings. The mini costs more but is much more versatile and necessarily embodies a powerful system for display and control of its functions.

Most of the 16-bit CPUs that have been produced have been intended and suited for inclusion as CPUs in minicomputers and, in this role, they have proved very successful. They can be used advantageously too in dedicated roles where their greater processing power has been found necessary and in this kind of application they have been cost effective—and cost effective in this context is the important criterion.

As the technology advances it is becoming as easy to produce 16-bit devices on a chip as it was to produce 8-bit devices previously. It should not be supposed that this will necessarily force the 8-bit device into obsolescence. Data is commonly handled in 8-bit bytes and this is likely to remain so for the immediate future. Thus, except when there is a demand for greater computing power, the 8-bit system seems likely to remain popular. It is interesting to note in this connection that a number of systems exist which are a mix of 8-bit and 16-bit devices. The 8-bit CPU handles the data and controls the system while 16-bit CPUs are kept busy with calculation.

From a philosophical point of view the development of 16-bit CPUs is of interest since it has meant that the design of minicomputers has become a preoccupation of many more designers from less restricted backgrounds. Computer design has always suffered from an incestuous restriction since so few people have been able to engage in it and the later generations have been inclined to follow their forebears rather uncritically. That this should be the case is a consequence of two main factors. The first reason is that as transistor device technology has developed, computers have gained in power merely from the increase in the raw logic speed of their components and until recently there has been little commercial compulsion for commercial designers to change the main features of their machines in order to produce improved capability. The second reason is, of course, the high cost of computer development. In a hard commercial environment, financial management has not been inclined to invest in experimental ideas that have any great risk of failure. The cheapness of the microprocessor will certainly remove some of this restriction from system development. Great numbers of people including amateurs and hobbyists will be able to afford to experiment. Doubtless there will be many failures but also many successful original ideas should evolve.

A typical rather original approach is exemplified by the Texas designers in their powerful TMS 9900 series minicomputer systems. Texas have historically been among the leaders in device design and have supplied a substantial number of the devices used by the computer industry. With the advent of LSI, like several other device manufacturers, they have determinedly entered the computer system field. The 9900 has an unusual contruction in that it exploits the concept that microcircuit RAM stores are little different from arrays of registers of the kind that normally form the pointers and central registers in the conventional microprocessor CPU. All the computing operations are performed 'store to store' instead of in the central registers. There are twenty registers on the chip that correspond to index and central registers but are in fact pointers to them in RAM store. This arrangement greatly facilitates context switching, that is, switching from one program to another which merely entails changing the set of pointers which can be done quickly and economically.

Several of the modern 16-bit CPUs are fabricated using the more complex but very much faster bipolar technology. These form the CPUs of 'single-board' computers which are necessarily minis because of their physical dimensions but have greater speed and power than the large mainframe machines of the previous generation. It is not unusual to hear of them rated at several 'Atlas powers', meaning that they have several times the capability of Kilburn's mighty Atlas of the 1960s.

BIT-SLICE MICROPROCESSORS

The easy way to conceive of these devices is to realise that a 16-bit CPU could fairly easily be arranged by juxtaposing two 8-bit CPUs in parallel—or four 4-bit devices. The true bit-slice device is, as it were, a slice 1 bit wide through a whole CPU so that a CPU of any width can be assembled by placing the required number of slices, or chips, side by side in parallel. One result of this is that it is possible to organise fairly simple array-computing devices hundreds of bits wide for special kinds of calculation needing this kind of precision. Array machines have been built for experimental purposes of up to 1024 bits width.

This bit-slice concept has become somewhat broadened to include a range of devices which are less vertical slices through the CPU but more the LSI provision of the organs of a computer on single chips. Thus chips are fabricated to be ALUs, PC and sequencing units, microprogramming units, and the like, which allow designers to build up conventional,—or unconventional—computer structures of whatever size and configuration they need but of only a few separate components. Normally, bit-slice processors are manufactured in bipolar technology and can be very much faster than MOS devices, which is one reason for this use.

And Even More Bits?

Hewlett-Packard and Intel have both recently announced 32-bit microprocessors. The former is to be used only within Hewlett Packard products. The Intel, called the 432, is for general marketing; it is to consist of three chips, presumably each with a pin-out of 40. Intel reckon that it will have the power of a medium to large mainframe such as an IBM 370/158. Certainly it would be wise to listen to what they claim: they were the first in the field and are still making the pace.

With a technology moving as rapidly as microelectronics it is hard to see where things are going. At present the great majority of potential users are still struggling to make the best of 8 or 16-bit machines. It remains to be seen who will find worthy uses for micros with mainframe power.

WHAT OF THE FUTURE?

The microprocessor as portrayed by the news media seemed to arrive with a bang and quite suddenly provided a great deal of sensation and many bones of contention for politicians, sociologists and others who regrettably know all too little about it. The magic chip caused consternation and dread among labour leaders since it 'would cause massive unemployment and have an appalling social impact'. It hardly seems to deserve so much heated discussion or such apprehension.

As we can see it is not the microprocessor itself which is central to all this sensation but simply the fact of LSI. The microprocessor is merely a very cheap computer and computers have been with us some twenty years or more. Many of us may remember that computers themselves once caused all kinds of prophecies of doom but became of less and less popular and media interest as they became more widely used and generally accepted. It seems likely that the microprocessor too will just become an accepted part of normal life. It does provide great scope for making much of life easier and more convenient. It seems likely, too, to provide great hosts of technically minded people with an interesting and intriguing challenge to find new and better ways of using it to make a better life and, of course, the converse—but that will be the fault of people not chips. An essential safeguard against its abuse will be the understanding of the microprocessor. It is the author's hope that this book has given the reader a good start.

Appendix A: Binary Arithmetic

NUMBER REPRESENTATION

Decimal numbers are based on radix 10. We express them in a structured form

$$C_n 10^n + \ldots + C_2 10^2 + C_1 10^1 + C_0 10^0 + C_{-1} 10^{-1} + \ldots + C_{-m} 10^{-m}$$

where each C_i is a coefficient chosen from the set of ten digits $\{0, 1, 2, 3, 4, 5, 6, 7, 8, 9\}$ and the indices show the power of 10 for each term. Thus the number 573.42 means

$$5 \times 10^2 + 7 \times 10^1 + 3 \times 10^0 + 4 \times 10^{-1} + 2 \times 10^{-2} \text{ or}$$

5 hundreds + 7 tens + 3 ones + 4 tenths + 2 hundredths. Any number of any magnitude, to any precision, can be expressed in this notation.

Ten was almost certainly taken as the radix or base because we have ten fingers to count on. It was an arbitrary choice: any number can be taken as a base although some would be more convenient to use than others. When numbers are to be dealt with in digital logic circuits it is most convenient to express them to the base 2, that is, in the binary system.

Numbers in binary are structured just as in decimal:

$$C_n 2^n + \ldots + C_2 2^2 + C_1 2^1 + C_0 2^0 + C_{-1} 2^{-1} + \ldots + C_{-m} 2^{-m}$$

but where each coefficient C_i is chosen from the set of two digits $\{0, 1\}$ and the indices show the power of two for each term. Thus the number 110.101 means

$$1 \times 2^2 + 1 \times 2^1 + 0 \times 2^0 + 1 \times 2^{-1} + 0 \times 2^{-2} + 1 \times 2^{-3} \text{ or}$$

1 four + 1 two + zero ones + 1 half + zero quarters + 1 eighth.

Any number of any magnitude or precision can be expressed in this form, also. The binary number shown is equal to 6⅝ or 6.625 to the base ten. To avoid any confusion we can indicate the base by a subscript, thus

$$110.101_2 \equiv 6.625_{10}$$

The decimal number has the decimal point to indicate the separation between integers and fractions: The binary number has the binary point as a separator. We should note that the binary fractions are the traditional fractions of the carpenter's rule.

From a human point of view binary numbers are inconvenient since any value expressed in binary needs about three times as many digits as it does in decimal. For instance a thousand in decimal is 10^3 or 1000_{10}. A near equivalent in binary is 2^{10} or 1024_{10} and is written in binary as 10000000000. It is not difficult to work out that the exact binary equivalent of 1000_{10} is 1111101000_2.

There are three pseudo-binary notations used by computer people to make life easier.

Octal notation

This simple notation merely takes advantage of the fact that each group of three binary digits represents the set of numbers from 0 to 7. Thus the number for decimal 1000 above, in octal is expressed as 1750_8 since this is interpreted as 1 | 111| 101 | 000 |, where each octal digit is expressed as its binary equivalent. The binary number is partitioned into groups of three digits from the binary point and these groups are written with the octal digits 0 to 7. Likewise for fractions 101·1101 is written in octal as 5.64, that is, | 5 | . | 6 | 4. Note that the least significant digit in the example number represents 100_2 or 4_8.

Octal representation has the advantage that it uses only eight ciphers 0 to 7 which are in the decimal digit range. It is three times more brief than pure binary and four or five digit numbers are easy to manipulate and remember.

Hexadecimal Representation (Hex)

In this representation the binary number is partitioned as in octal but with four digits to a partition. Thus the range of decimal numbers in each partition is 0 to 15. Since there are no decimal digits of greater value than 9 we use the first six letters of the alphabet to represent them

$$A = 10$$
$$B = 11$$
$$C = 12$$
$$D = 13$$
$$E = 14$$
$$F = 15$$

Thus the hexadecimal or hex number E7.B$_{16}$ 1110 | 0111| . | 1011 or
11100111.1011$_2$. The decimal equivalent is easy to derive

$$(14 \times 16) + 7 + {}^{11}/_{16}$$

which is $224 + 7 + (^{11}/_{16})$ or 231.6875_{10}.

The hex notation is nicely compressed. It is somewhat difficult to manipulate without practice, but it fits nicely into 8-bit bytes, since each requires two symbols which are available in any normal typeface.

Binary Coded Decimal (BCD)

Anyone who has experimented with conversion between binary and decimal and vice versa will be quite familiar with the difficulties it entails. The conversion is not difficult to program for a computer but there still remain difficulties of exactness especially in dealing with commercial transactions where 'to the nearest penny' is not an acceptable approximation on a balance sheet. To avoid this difficulty many business machines, like pocket calculators, operate in BCD for their arithmetic processes connected with the problem itself. They still work in pure binary for all their internal management activities such as addressing and indexing.

In BCD notation the ten decimal ciphers or digits are represented by their 4-bit binary equivalents. Thus the number 93.46 is expressed as 1001 0011 . 0100 0110 or compressed to 10010011.01000110. In fact within the computer registers, the binary point is itself not registered. Its position is fixed by the position of the numbers in the register and by the arithmetic rules by which the device works. Typically in pocket calculators, the decimal point is displayed and manipulated by separate arithmetic from that which does the numerical operations. The conventions for doing this will be quite familiar to anybody who has used the old fashioned mechanical calculating machines.

In BCD the binary equivalents of decimal 10 to 15 are not used. When BCD arithmetic is performed in conventional binary arithmetic units the occurrence of these numbers must be detected and the values adjusted. Most microprocessors provide a function known as 'decimal adjust' to facilitate this.

BINARY ARITHMETIC

The operations of binary arithmetic follow the same rules as decimal arithmetic but are simpler. For instance the addition and multiplication tables for binary arithmetic may be summarised as

Add	0	1
0	0	1
1	1	10

Multiply	0	1
0	0	0
1	0	1

or in symbols

and

$0 + 0 = 0$
$0 + 1 = 1$ addition
$1 + 1 = 10$
$0 \times 0 = 0$
$0 \times 1 = 0$ multiplication
$1 \times 1 = 1$

Simple examples should suffice to demonstrate the operations on multi-digit numbers.

Addition

$$\begin{array}{rl} 1\,0\,1\,1 & = 11_{10} \\ 1\,1\,0\,1 & = 13_{10} \\ \hline 1\,1\,0\,0\,0 & = 24_{10} \end{array}$$

Multiplication

$$\begin{array}{rl} 1\,0\,1\,1 & = 11_{10} \\ 1\,1\,0\,1 & = 13_{10} \\ \hline 1\,0\,1\,1 & \\ 0\,0\,0\,0\,0 & \\ 1\,0\,1\,1\,0\,0 & \\ 1\,0\,1\,1\,0\,0\,0 & \\ \hline 1\,0\,0\,0\,1\,1\,1\,1 & = 143_{10} \end{array}$$

Subtraction

$$\begin{array}{rl} 1\,1\,0\,1 & = 13_{10} \\ 1\,0\,1\,1 & = 11_{10} \\ \hline 0\,0\,1\,0 & = 2_{10} \end{array}$$

Division

$$\begin{array}{r} 1\,1\,0\,1 \\ 1\,0\,1\,1\,\big|\,\overline{1\,0\,0\,0\,1\,1\,1\,1} \\ 0\,1\,0\,1\,1 \\ \hline 0\,0\,1\,1\,0\,1 \\ 0\,0\,1\,0\,1\,1 \\ \hline 0\,0\,0\,0\,1\,0\,1\,1 \\ 0\,0\,0\,0\,1\,0\,1\,1 \\ \hline \end{array}$$

It should be noted that the procedure is exactly as for decimal arithmetic but from the point of view of mechanisation it is simple indeed. For instance the multiplication procedure is

Appendix A: Binary Arithmetic

(1) if the least significant digit of the multiplier is 1, copy the multiplicand to form the first partial total. If the least significant digit is 0, do nothing

(2) if the next least significant digit is 1, do as (1) above but shift the copied value one place left

and so on. The whole operation involves only adding, left shifting and copying. To make the addition process easier, in a computer the multiplication procedure becomes

```
         1 0 1 1
         1 1 0 1
         -------
         1 0 1 1
       0 0 0 0 0
       ---------
       0 0 1 0 1 1
       1 0 1 1 0 0
       -----------
       1 1 0 1 1 1
     1 0 1 1 0 0 0
     -------------
     1 0 0 0 1 1 1 1
```

In this arrangement no summation exceeds $1 + 1 +$ a carried 1, or $1 + 1 + 1$, the sum of 3 digits.

Simple arithmetic units such as are embodied in microprocessors can only add and sometimes subtract. Multiplication is performed by program, as is division. Larger computer ALUs include multiplication but rarely division. Division poses problems due to decisions having to be made at every step as to whether to shift or not. There are numerous artifices but common practice is to program division using a non-restoring technique or to derive the reciprocal of the divisor and multiply the dividend by it. There is a wealth of information about these techniques in programming manuals and books on digital calculation methods. They are not important at this level of study.

The precision of numbers which can be dealt with in single byte operations is trivial for practical calculations because of the short word length of the 8-bit microprocessor although it is adequate for store addressing purposes. For this reason higher precision arithmetic is usually performed in BCD by program, following the rules of decimal arithmetic and using decimal multiplication tables held in store.

SIGN CONVENTION

There is one important aspect of binary arithmetic that is important in the context of microprocessors: in the course of even simple operations negative numbers may occur. There must not only be a way of indicating the sign of numbers; a convention must be established by which negative numbers occurring in the course of calculation can be detected as such by the ALU logic and dealt with accordingly.

The standard approach is as follows. All numbers are scaled to be binary fractions and all manipulations are carried out in fractional mode. All negative numbers are represented by their complements modulo 2. Simply, what this means is that any positive number is divided by a scaling factor until it lies in the range $0 \leq n < 1$: a negative number, also within the same range is subtracted from 2. It then must be in the range $1 \leq n < 2$.

Using this convention, in an 8-bit register, all negative numbers have a 1 as their most significant digit, while all positive numbers have a zero. The binary point becomes fixed in the position immediately following the most significant digit. Some simple examples should suffice to illustrate the working of this convention.

$$0 1 1 0 0 0 0 0 = \tfrac{1}{2} + \tfrac{1}{4} = \tfrac{3}{4}$$

To determine the representation of $-\tfrac{3}{4}$ (the binary point is shown here for clarity)

	1 0 . 0 0 0 0 0 0 0	2
Subtract	0 . 1 1 0 0 0 0 0	$\tfrac{3}{4}$
	1 . 0 1 0 0 0 0 0	$-\tfrac{3}{4}$ (actually $1\tfrac{1}{4}$)

By the nature of the logic of an ALU, the 1 on the left of the upper number, 2, is assumed to be present but is out of the range of the register. Thus subtracting $\tfrac{3}{4}$ from 0 in the machine achieves the identical result.

Subtraction may be provided as a logical function but in many ALUs the process is performed by negation of the subtractor and adding. Thus $\tfrac{7}{8} - \tfrac{3}{4}$ is achieved by $\tfrac{7}{8} + (-\tfrac{3}{4})$

	0 . 1 1 1 0 0 0 0	$\tfrac{7}{8}$
Add	1 . 0 1 0 0 0 0 0	$-\tfrac{3}{4}$
	0 0 0 1 0 0 0 0	$\tfrac{1}{8}$

Observe also ($\tfrac{1}{8} - \tfrac{7}{8}$) using subtraction

	0 . 0 0 1 0 0 0 0	$\tfrac{1}{8}$
Subtract	0 . 1 1 1 0 0 0 0	$\tfrac{7}{8}$
	1 . 0 1 0 0 0 0 0	$-\tfrac{3}{4}$

Here the 1 to the left of the most significant register has been assumed by the logic of the subtractor unit.

THE NEGATION PROCESS

The complement modulo 2, or negative representation shown above was derived by subtraction from 0, or more truly 2. It can easily be established with some simple examples that the negative representation of a positive number within the permitted range can be directly written down by complementing each digit, except the least significant one, starting from the left. This is easy for a human but less so for a logic unit like an ALU. The same result can be obtained, however, in a way that is entirely convenient to an ALU. First, every digit is complemented, including non-significant zeros. Second, 1 is added in the least significant position. For example

$$
\begin{array}{rl}
0.1100000 & = 3/4 \\
1.0011111 & = \text{one's complement of } 3/4 \\
\text{Add } 0.0000001 & \\
\hline
1.0100000 & = -3/4
\end{array}
$$

The one's complement method of negative number representation is sometimes used in computers, but the complement modulo 2 is generally preferred.

There is a great deal more to binary arithmetic than has been demonstrated here. Anybody who becomes seriously involved with data processing, particularly with microprocessors, will get ample practice at it. There are many excellent expositions of it in maths and computer books. This brief introduction should, however, suffice for the understanding of simple microprocessor operation.

Appendix B: Computer Logic

INTRODUCTION

The logic of computers and digital systems is called *combinational logic*; it is a direct derivative of Boolean algebra and is often called that, although Boolean algebra is, in fact, a rigorous mathematical discipline and is much more comprehensive than combinational logic. Boolean algebra was evolved by the mathematician George Boole in the 19th century to establish the validity or otherwise of deductions or arguments. Boole's logic was mostly concerned with the truth or falsehood of combinations of propositions which themselves had the sole property of being either true or false, in other words, binary valued. The engineer's combinational logic is also primarily concerned with combinations of binary valued quantities or variables and hence its name.

It is easier to make electronic devices that have two unambiguous states than it is to make devices that have several. For this reason computer designers have developed binary digital systems in preference to others. Had the earliest designers seen computers as logical rather than calculating devices they might well have formed the same preference anyway. In fact some decimal systems have been built as well as a few experimental ternary (three-state) systems.

Logic elements, that is, devices that implement or model the rules of logic, can be mechanical, electrical, pneumatic or electronic, valve or transistor. One of the advantages of studying computers as logic systems is that the concepts of their operation are independent of the means of their implementation. In the study of microprocessors we are concerned only with transistor-based elements.

LOGIC ELEMENTS

Digital systems today are implemented exclusively with two-state or binary devices, for instance, switches, that can be on or off, bistables that can be set or reset and circuits whose inputs and outputs can be high or low, conducting or non-conducting. In digital circuits it is normal to adopt conventions ensuring that the two permitted states can be unambiguously distinguished. A typical convention is that high or true or logic 1 is nominally 5 V, while low or false or logic 0 is 0 V. To ensure that the real circuit states are unambiguous in operational conditions, a value between 2.5 and 5 V is taken as high while a value below 0.5 V is taken as low. A value occurring in between these two ranges is taken as indicating a circuit fault. The gap between them is called the 'noise margin', since an intruding noise voltage added or subtracted from a valid signal needs to be as large as the gap voltage to cause an erroneous value to occur. Logic elements are designed to have amplification so that on receipt of inputs within the permitted ranges, their outputs are at, or close to the nominal values.

Listed below are the proporties and symbols of the commonly used logic elements.

AND

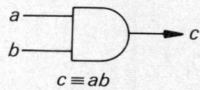

Figure B.1

The output c is true if, and only if, all the inputs are true. There may be more than the two inputs that are shown in figure B.1.

OR

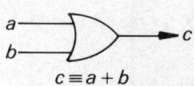

Figure B.2

The output c is true if any of the inputs are true. See figure B.2.

NOT

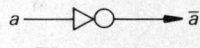

Figure B.3

The output is false if the input is true and the output is true if the input is false. See figure B.3.

NAND

Figure B.4

The output is false if, and only if, all the inputs are true, or the output is true if any input is false. See figure B.4.

NOR

$$c \equiv \overline{a+b}$$

Figure B.5

The output is true if, and only if, no input is true, or the output is false if any input is true. See figure B.5.

EXCLUSIVE OR

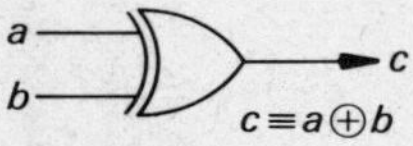

$$c \equiv a \oplus b$$

Figure B.6

The output is true if either input is true but not both; or the output is true if the inputs differ. See figure B.6.

These properties can be summarised as a truth table.

Inputs		Outputs					
a	b	AND	OR	NOT a	NAND	NOR	$a \oplus b$
F	F	F	F	T	T	T	F
F	T	F	T	T	T	F	T
T	F	F	T	F	T	F	T
T	T	T	T	F	F	F	F

COMBINATIONAL LOGIC

The symbols used in combinational logic are few and have been chosen so that they can be typed. The symbol for AND is '·' and is normally omitted. Thus 'a AND b' is written '$a \cdot b$' or just 'ab'. The symbol for OR is '+' and this must not be confused with the plus of arithmetic addition. The normal notation for true and false used by circuit designers is 1 and 0. Note that 1 + 1 in logic, meaning true OR true, means only 'true' and not 'twice true'; this figure '1' means unity or integrity or one hundred per cent true. There need be very few occasions when the logical and arithmetic notations can be confused.

The NOT operation is signified by a bar over the variable or expression that is operated on. It is also called 'inversion' or 'complementation'. Thus we say the complement of a is NOT a, written as $\overline{a}$. The complement of $a + b$ is NOT $(a + b)$, written $\overline{a + b}$. Likewise the complement of ab is written $\overline{ab}$ and we should note that this is not the same as $\overline{a}\,\overline{b}$.

Appendix B: Computer Logic

There are a small number of theorems used in combinational logic; these are tabulated below.

(1) $0 \cdot a = 0$; $1 + a = 1$
(2) $1 \cdot a = a$; $0 + a = a$
(3) $aa = a$; $a + a = a$
(4) $a\bar{a} = 0$; $a + \bar{a} = 1$
(5) $ab = ba$; $a + b = b + a$
(6) $abc = a(bc) = (ab)c$; $a + b + c = (a + b) + c = a + (b + c)$
(7) $\overline{abc} = \bar{a} + \bar{b} + \bar{c}$; $\overline{a + b + c} = \overline{abc}$
(8) $ab + ac = a(b+c)$; $(a + b)(a + c) = a + bc$
(9) $ab + a\bar{b} = a$; $(a + \bar{b})(a + b) = a$
(10) $a + ab = a$; $a(a + b) = a$
(11) $a + \bar{a}b = a + b$; $\bar{a}(a + b) = ab$
(12) $ab + \bar{a}c = ab + \bar{a}c + bc$; $(a + b)(\bar{a} + c) = (a + b)(\bar{a} + c)(b + c)$

The theorems are quite simple: they are mostly the kind of commonsense conclusions that are not so easy to prove by a rigorous mathematical or logical proof. There is, however, a useful graphical method of demonstrating them which also provides a valuable aid to logic design. Those readers already schooled in symbolic logic will recognise the technique as a direct descendant of Venn's and Marquand's diagrams.

THE KARNAUGH OR K MAP

Suppose we draw a square as in (i) below:

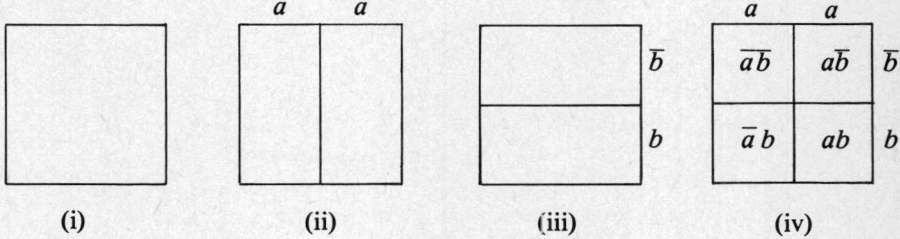

(i) (ii) (iii) (iv)

This square represents the logician's 'universe of argument'. Consider a proposition or 'variable', a; if we then bisect the square as in (ii) we can say that the right hand half of the square represents the field of a being true; and so, the other half must represent the field of a not true, if a is binary valued. In which case we can label this half $\bar{a}$. Suppose we now take a second binary valued proposition within the same universe. We can draw the square (iii), bisect it and label it as shown. By combining the squares we can produce the square (iv), which also combines the propositions or variables.

For a trivial example, consider the universe of argument to be human beings. We take a as the proposition that a human is male, and b that a human has blue eyes. Then in the square (ii) $\bar{a}$ is the field of non-male humans, presumably, females. Likewise $\bar{b}$ will denote the field of humans who have eyes any other colour than blue. Then ab is the field of blue-eyed males. Those humans who are not blue-eyed males must be either those who have not blue eyes OR those who are not male; that is the other three squares, covered by the two halves $\bar{a}$ and $\bar{b}$.

Using the squares in the same way we can demonstrate the truth of De Morgan's theorems in the list above; $\overline{ab}$, that is, NOT(a AND b) is the area given by $\bar{a}$ OR $\bar{b}$. Likewise $a + b$, that is, a OR b covers three squares. Its complement $\overline{a + b}$ is the remaining square $\bar{a}\bar{b}$.

The K map can easily be expanded to deal with three or four variables, as follows

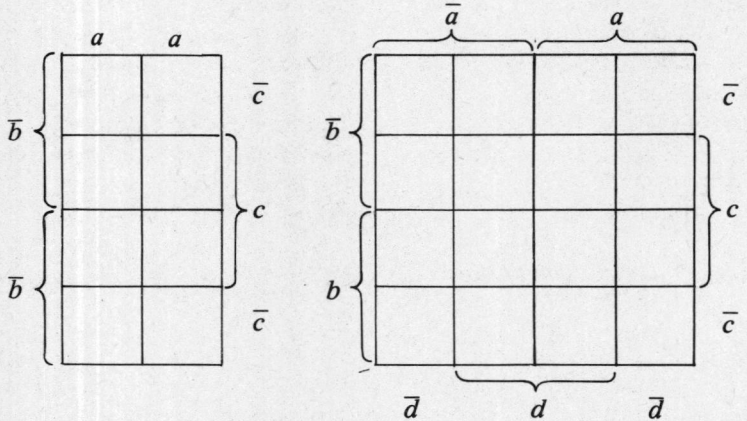

It can be expanded a little less easily to cope with five or six variables and very much less easily to cope with even more.

The map technique is a very useful aid in simple logic design. It provides a simple way of determining complements of Boolean functions and of developing minimal expressions, that is, expressions in their simplest and shortest form. Complementation is continuously required when dealing with NAND or NOR gates and minimisation leads to economical design.

There are a multitude of books available which deal with all levels of detail of logic design. This brief exposé has been limited to that detail required to understand microprocessors and their systems and, hopefully, to whet the appetite of some readers to investigate the subject more fully.

SEQUENTIAL LOGIC ELEMENTS

There is another kind of logic elements that are extensively used in microprocessors and digital systems. These elements are called sequential logic elements. They have the quality that their output depends, at any instant, not only on their input but also on the previous inputs.

Suppose we interconnect two NAND elements as shown in figure B.7.

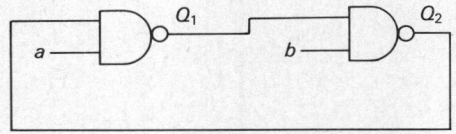

Figure B.7

Suppose we hold input a to logic 0, then we know from the definition of the NAND gate that its output Q_1 must be 1. If we hold b to 1, then the second NAND gate has as inputs $b = Q_1 = 1$ and its output must, by definition, be 0. Since this is so, we can change a to 1 without altering the state of the circuit. The circuit is, however, symmetrical. Instead of a, we could have made b to have the value 0. In this case Q_2 must equal 1 and if $a = 1$ then $Q_1 = 0$. Either state of the circuit is stable so long as $a = b = 1$, and we therefore have a bistable circuit formed of two NAND elements. We can redraw the circuit as shown in figure B.8.

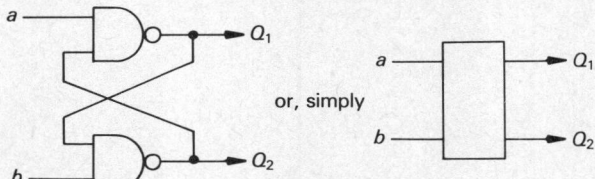

Figure B.8

We should notice that so long as a and b are never both 0 simultaneously $Q_1 = \overline{Q_2}$, that is, the outputs are complementary. We can improve this circuit quite simply using the diagram in figure B.9. We now have a gated bistable, or simple latch. If c is held to 0, both input gates output 1s and the bistable can be in either of its two states. If we

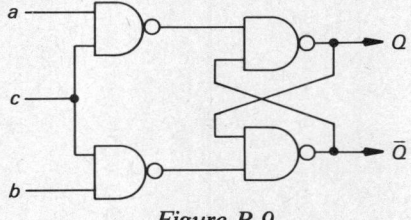

Figure B.9

now apply a 1 to c, called the 'control' or 'clock' input, then the input gates are 'enabled': so long as $a = b = 0$ nothing happens. If a is set to 1, Q goes to 1 if it is not already in that state. We call this the 'set' state and the input a, the 'set input'. Conversely if we had set b to 1, $\overline{Q}$ would have become 1 and the bistable is in the reset state; the input b is the reset input.

A further addition of an inverter, or NOT element, gives us a 'D' type latch shown in figure B.10, where D presumably stands for 'data'.

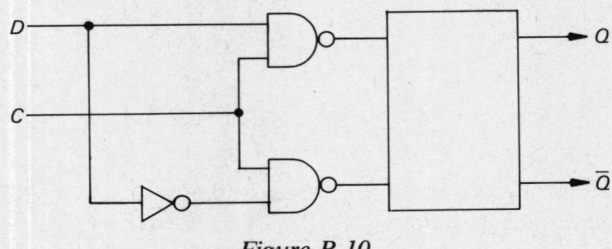

Figure B.10

This element forms the simplest stage for building a register as shown in figure B.11.

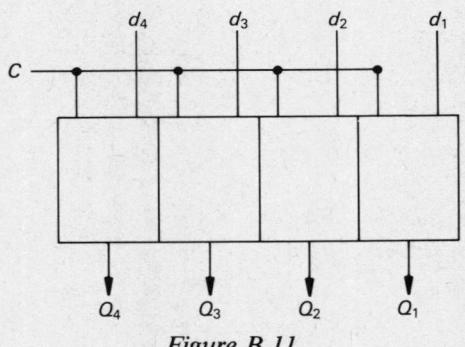

Figure B.11

When C is held at 1 the data at d_1 to d_4 sets or resets the bistables to the current input values. When C is taken to 0, the values of d_1 to d_4 at that instant are latched into the bistables and displayed by the states of Q_1 to Q_4. Note carefully, however, so long as C is held to 1 the outputs Q merely follow the inputs d. The output is only 'staticised' at the instant that C changes from 1 to 0.

D-type latch registers are the simplest to implement but they cannot be used in the implementation of arithmetic operations like $A \leftarrow A + B$, as in an accumulator process. A is the output of the accumulator: if it is added to B and then fed into the input again an indeterminate result will follow, since there is no separation between output and input. In this case we need two registers for A or two levels of bistables—this is shown

Appendix B: Computer Logic

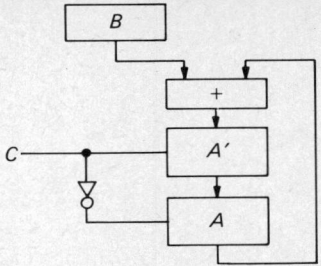

Figure B.12

in figure B.12. Here the contents of A and B are added and the arithmetic sum $A + B$ is applied to the input of A^1. When C is 1, A^1 assumes the value $A + B$, but A is not affected since its control input is at 0. When C goes to 0 the input to A^1 is disabled, but its output can set $A + B$ into A as required. The registers in a microprocessor are necessarily as simple as possible and this is a principal reason why the ALU was shown in the CPU diagram in chapter 2 as having auxiliary registers. The technique of using two levels of latched bistables is called the 'master/slave' technique. There are several individual bistable circuits that are based on it.

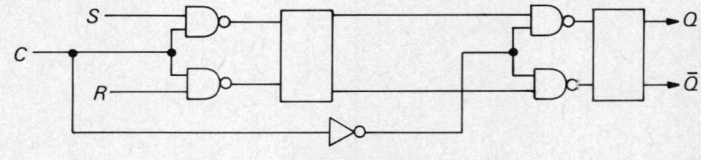

Figure B.13

The bistable shown in figure B.13 consists of two gated latches with complementary C inputs, as in the register arrangement described above. Now there is no direct path from S and R to Q and $\overline{Q}$. When C goes to 1 the master bistable is set to the value dictated by S and R. Only when C goes down, gating out S and R, does the slave bistable assume the original S and R values.

Now consider the circuit shown in figure B.13 but with cross-coupled feedback. This circuit is shown in figure B.14. Suppose $Q = 1$, $\overline{Q} = 0$ and $C = 0$. Then the feedback line from Q puts 1 on the R input of the master bistable, while $\overline{Q}$ puts 0 on S. When C goes to 1 the master

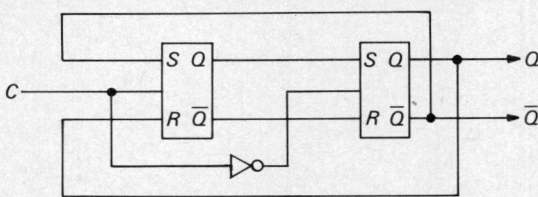

Figure B.14

must become reset. When C goes down to 0 again Q goes to 0 and $\overline{Q}$ to 1, that is, the slave has changed value or 'toggled'. If a succession of pulses is applied to C then this bistable will change over or toggle once for each complete pulse. Its output becomes a train of pulses but at half the repetition frequency of the pulses at C. It is thus a divide-by-two circuit. If its output is applied to a similar circuit, the output of this will be again a division of two. A chain of them becomes a binary counter; after a series of pulses applied to it the Q values will indicate the binary count of the applied pulses.

Note, however, the complexity of the circuit. As a rough estimate each NAND element comprises at least three transistors. Each gated bistable comprises at least four NAND gates and a counter stage, at least nine. For a 16-stage register a minimum transistor count will require 16 × 12 transistors, that is, 192. A 16-stage counter will require at least 16 × 9 × 3, which is 432 transistors.

There is one further bistable development which we should examine. It is called for historical reasons a J-K bistable. If we look back at the master/slave bistables described previously, we can combine them, that is, have three-input NAND elements as the set and reset control gates of the master bistable. This arrangement is shown in figure B.15. The inputs

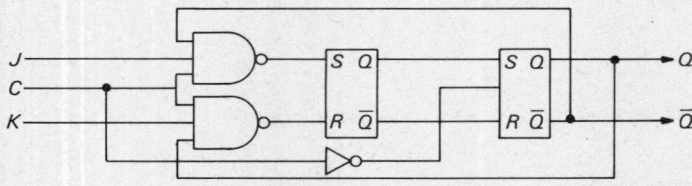

Figure B.15

J, K have been added to the input gates. Referring back to the definitions we appreciate that while $J = K = 1$, the circuit is unchanged from the counter bistable. If we pick our way carefully through the circuit we shall find also that if $J = K = 0$, then the circuit ignores the C input: However if $J \neq K$, Q is set by J, or $\overline{Q}$ by K, as the C value is taken to 1 and back to 0, that is, when C is supplied with a pulse.

Thus the $J-K$ bistable is versatile: it can be used as a register stage, as a shifting register stage or as a counter stage. It is widely used in logic design and is an important member of every family of logic elements. It is also complicated and its use is necessarily very restricted inside microprocessors.

THE LOGIC OF AN ALU

Although the design of ALUs and arithmetic microcircuits is a highly specialised art, an examination of ALU logic is a useful aid to the

Appendix B: Computer Logic

understanding of both logic and microprocessors. Consider first the process of addition of binary numbers; for example

$$
\begin{array}{ll}
101 \ (= \ 5) & (a) \\
\underline{110} \ (= \ 6) & (b) \\
1011 \ (= 11) & (a + b)
\end{array}
$$

Doing the sum in exactly the way we would do a decimal addition, we first take the least significant or right hand digit of each number and add them: $1 + 0 = 1$, so we can write down the rightmost digit of the resultant. We take the next digit of each number and we have a similar result, note that in neither case has a carry occurred. The third digits of the two numbers are both 1, and $1 + 1 = 10_2$ (decimal 2). So we write down 0 as the resultant of the third pair of digits and have 1 to carry. Since the next pair are 0s, we have $1 + 0 + 0 = 1$ and enter 1 as the result.

In order to implement the process of arithmetic we are going to use logic elements. We must therefore reduce the arithmetic process to a logic process. In the first place we need to register the numbers we want to add, so an array of bistables will be required for each register. Since we need to set them to the required values, the bistables can conveniently be latches. The addition process is then to apply the output of the two end latches as inputs to a logic element which, by comparing them, generates a 1 or 0 according to the rules of arithmetic. Obviously binary arithmetic lends itself to this: decimal arithmetic would be much more difficult. The logical requirement for the binary addition value to be generated can be summarised: if both digit inputs are the same, generate a 0; if the digit inputs differ, generate a 1. Reference to the earlier tabulated list of logic elements shows that this requirement is met by an exclusive OR or anticoincidence element, often referred to as a 'half-adder' because it has this capability.

After the first pair of digits, however, there may be a carry and this will need to be considered when dealing with the next and successive stages of the addition. Reverting again to logic: whatever value the two number digits generate, if we compare it with the carry generated by the previous addition, we find that if the carry is 0, the result is correct. If the carry is 1 the result will have to be complemented. If we tabulate the possibilities as a truth table we find that, again, the anticoincidence element will perform the process. Also we see that two half-adders give a full adder, that is the complete adder logic for a stage with carry. Naturally, the carry itself must be generated: if both digits of the first stage are 1, then the carry is needed and an AND element will perform the operation. It is not quite so simple for the succeeding stages. A carry must be generated if at least two out of three of the inputs is a 1 at each stage. Notice, however, that we have reduced the arithmetic numerical problem to a purely logical problem and we could equally well be writing true or false for 1 or 0.

Suppose we try a more systematic approach. Suppose we call the digit values at each stage a and b, and the carry from the previous stage, c. We are concerned to find, at each stage, the right binary values to represent the sum digit d and the new carry c_n. It is not difficult to write down a truth table for d and c_n with a, b, c as inputs. We can reason as follows: d should be 1 (or true) if one and only one of a, b, c are 1 (or true) or if all three are 1 (or true). Likewise, c_n should be 1 if two or three of the a, b, c values are 1 but not otherwise. We can write these conditions down as Boolean expressions

$$d \equiv a\bar{b}\bar{c} + \bar{a}b\bar{c} + \bar{a}\bar{b}c + abc$$
$$c_n \equiv ab\bar{c} + a\bar{b}c + \bar{a}bc + abc$$

We must remember that '+' here is 'OR': thus d is true if a is true and b and c false OR if a and c are false and b is true OR ... etc. It is interesting and profitable to write these expressions into Karnaugh maps. It immediately becomes apparent that although the expression for d is the best expression we can get, c_n can be written more simply as $c_n \equiv ab + bc + ca$, which is manifestly a saving when it comes to implementation with hardware elements. These expressions for d and c_n provide the complete logical design for an addition stage when they are implemented. Thus an 8-bit ALU such as that in a microprocessor would have eight such networks arranged in a parallel array. The carry from the most significant bit position would be stored in a bistable in the flag register, showing whether a carry or overflow has occurred during the addition process.

Another illustrative example of logical design and implementation in ALUs is that of the 'decimal adjust' function. This is used when the ALU is operated in binary coded decimal BCD mode. In this mode decimal digits are encoded as their binary equivalents using 4 binary digits. Now 4 binary digits are capable of registering or signifying 16 combinations; those signifying 10 to 15 are not required: if they occur in the course of addition then an adjustment must be made so that, for instance, 12 is converted to 2 and the excess 10 is carried to the digit group of next higher significance. Since this adjustment must be carried out automatically if the adjust function is invoked, a logic network is required that can detect the presence of any of the 6 excess numbers at the end of an add operation. Examination of the excess numbers leads to the following conclusions

(1) they are all greater than 8, thus the most significant bit is always a '1'.
(2) the least significant bit can be 1 or 0 and does not need to be considered.
(3) either or both the middle two bits must be 1.

The logical conditions therefore for the adjustment to be necessary are as follows. Adjust if the most significant bit = 1 AND if bits 2 OR 3 are 1, that is

$$(\text{bit 4}) \text{AND}(\text{bit 2 OR bit 3}) \rightarrow \text{adjust}.$$

If the reader chooses to experiment with the figures it will be found that the actual adjustment process, if required, can be achieved by subtracting 6 or adding 10 to the BCD sum.

These two examples of logic design are only illustrative. It has not been the author's intention to try and teach the whole complicated art of logic design in a few short paragraphs. It is hoped, though, that at least some of the readers who have followed so far may want to interest themselves in the subject more thoroughly and that they will find it amusing and valuable.

BUSBAR LOGIC

An important feature of microprocessors is their dependence on shared address and data buses or highways. These are of particular importance since, whereas the great majority of logic in the processors and ancillary units is locked away safely within the chips, these external highways are shared and necessarily available to outside intervention. As far as the processor is concerned, the address bus is unidirectional, outwards, and the data bus is bidirectional. Figure B.16 shows the typical arrangement of the highways and control signal buses. Like the data bus lines the control lines may be bidirectional in function, in some cases, although they are usually either inward to the processor or outgoing from it. The number of them will vary with the type of processor used and the type of application.

Logic devices, though tiny, do have a finite power consumption and draw a small amount of power at their input from the output of the device that signals to them. Generally the driving device is also tiny and is only capable of putting out enough power to operate one or two other elements. The power required to drive an element is called a logic load. Manufacturers of element families define the value of these logic loads and also define the number of loads that each logic element can drive. It is always necessary before trying to design a logic system for the designer to acquaint himself thoroughly with the loading rules for the elements he intends to use. It must never be forgotten, too, that at the speed of present day logic and with the low power available even a short length of wire or printed circuit track poses an appreciable load. The contact resistances and even contact potentials of plugs and sockets and

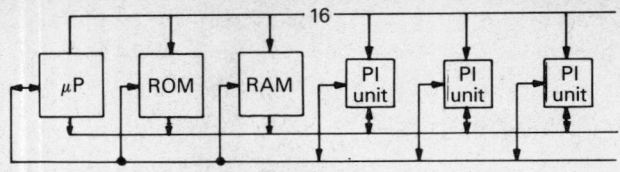

Figure B.16

connectors are not beneath consideration. To aid the designer to surmount the loading problem there are a variety of buffer elements available; these are power amplifiers, having an input load that is small, but are capable of driving a number of loads on their output. Sometimes they are logical inverters but usually they have no logic function. They do, however, necessarily cause delays, having switch on and off times comparable to those of the logic elements with which they are designed to work.

In busbar systems, many elements must necessarily be connected to the same bus line or point. Some connections may be inputs and some outputs but all impose loads on the element that is currently engaged in driving the line. There is a technique now available which considerably alleviates this problem of multiple loads: it is called 'tri-state' logic. Most current microprocessors and their ancillaries normally use this technique on their output terminals and several manufacturers have produced logic elements such as latches and buffers that also are tri-state. The meaning of tri-state in this context is simply that the device can output or accept a logic 1 or 0 level in the normal way but may also be switched into a high impedance state, that is, a condition in which it presents to the bus bar virtually no load at all. Thus if all the devices connected to a bus are tri-state only those which are involved in a data transfer effectively load the lines. By this means it is possible to interconnect a substantial number of devices to a busbar.

Another common technique of busbar operation is often called 'pull down logic' or 'open collector'. The principle of this is illustrated in figure B.17.

The output element of each device is a transistor, and normally it is held 'cut off', that is, in a non-conducting state. In this state it has little effect on the busbar line. If any of the transistors is caused to conduct

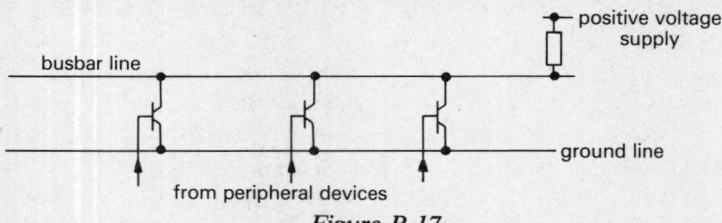

Figure B.17

then the voltage of the rail drops from high, that is, close to the positive supply voltage, to low, that is close to ground or zero voltage. It is the use of this technique which dictates the common use of negative or 'low' signal levels on microprocessor bus systems, leading to the frequent use of abbreviations like INTREQLOW; this means that the interrupt request signal causes the interrupt line to be pulled down from its at-rest, high voltage, to ground, or low.

These notes on busbar logic are not truly related to the logic but to the electronics of the system. However the implementation of the techniques is usually part of each logic element or the microprocessor.

Appendix C: Transistors and Microcircuits

HISTORICAL DEVELOPMENT

Transistors are the basic elements of all microcircuits. They can be broadly categorised into two types: junction transistors and field effect transistors. The junction transistor was used universally for a long time. For digital circuits it has the advantage of being a very fast switch. It is a complicated device to fabricate and the principles of its operation are not very easy to understand by anyone not acquainted with atomic physics. The more recently developed field effect transistor is of much simpler construction, takes very much less power and is much easier to understand. It is more commonly known today as a MOS or metal oxide silicon transistor. A great deal of research and development has gone into the technology of MOS devices and they now exist in a number of forms. They are the preferred devices in microprocessor manufacture except where speed is the prime consideration because they are simpler to fabricate and each element takes up much less space on a chip.

The junction transistors, used in what is called the bipolar technology, were originally made by a complicated process of 'growing' junctions between semiconductor alloys of different impurity concentrations; they were marketed as discrete transistors. The technology of making so called 'planar' transistors was then developed, that is, transistors formed on the surface of a thin silicon wafer by a series of photo masking and etching processes. They are fabricated, hundreds at a time, as arrays of elements on a single wafer. The wafer is then broken up into 'dice', each containing a transistor. The dice are tested and if satisfactory, encapsulated and marketed as discrete transistors. The percentage of good elements out of a fabrication batch is called the 'yield'. In the early days yields were low, until the fabricators gained skill and experience. When the yields became consistently high it became possible to combine and interconnect small groups of transistors on the wafer and form 'integrated circuits'. These groups are parted off the wafer as dice and encapsulated, forming the range of what are now called small scale integrated (SSI) circuits. Digital circuits of this kind are used extensively in computers and digital systems. A typical capsule is the dual-in-line (DIL) microcircuit logic element containing 2, 3 or 4 gates or two bistables.

Appendix C: Transistors and Microcircuits

The technology continued to advance and medium scale integrated (MSI) circuits have become commonplace. These elements embody hundreds of transistors per chip rather than the tens of SSI. Typical devices are 4-bit adder units, 4 or 8-bit registers and counters. The SSI chip is usually encapsulated in a 14 or 16-pin package. To accommodate the more complex circuits possible using MSI techniques, larger encapsulations have been developed with up to 24 terminals or pins.

The techniques of fabricating the field effect or MOS transistor began to become commercial more recently. These, too, first appeared as discrete devices: Their low power consumption and particular characteristics made them especially useful over a wide range of applications in the analog field. They were considered, at first, to be much too slow for digital system applications. As the technology improved the devices got faster and became embodied in SSI and MSI circuit configurations more or less analogous to the bipolar devices. They were, and are, of particular use where speed is less important than low power consumption.

The very small size and quite simple configuration of the MOS transistor made it possible and economical to fabricate thousands on a chip rather than hundreds and this led to the era of large scale integrated (LSI) circuits. The first major exploitation of these possibilities was the pocket calculator. This was followed by microcircuit storage elements and then by the microprocessor. We are now moving towards VLSI (very large scale integration), with tens of thousands of transistors on a chip.

SEMICONDUCTORS

The capabilities of all transistors depend on the physical and electrical attributes of a group of chemical elements called semi conductors. There are three of these which have been extensively used in electronics: selenium, germanium and silicon. Originally transistors were fabricated from germanium but the great preponderance of transistors today are made from silicon.

Semiconductors are so called because they can be both insulators and conductors according to how they are treated and used. They are all from the chemical group of quadri-valent elements, that is to say, their atoms each embody four valency electrons. Valency electrons are those which form the outer shell of atoms and are used to bond the atoms together as crystals or to bond them to other elements in chemical combination.

Silicon has the property of crystallising into a perfectly regular crystal lattice. The atoms can be envisaged in this consideration as being regular tetrahedrons, that is, three dimensional shapes, made up of four equal size equilateral triangles (similar to the cardboard milk cartons that used to be popular not so long ago). If molten silicon is allowed to crystallise on cooling in ideal conditions, and if it is absolutely pure (called 'intrinsically' pure) then it forms crystals with perfect lattices. In this form every bonding electron of every atom is totally engaged in holding the atoms together.

Materials that conduct electricity must have free electrons, that is, electrons that are free to migrate from atom to atom through the lattice and be charge carriers, under the influence of an applied electric field. Silicon crystals at very low temperatures have no free electrons and are therefore perfect insulators. With increased temperature the electron bonds become agitated. At ambient temperatures in silicon a quite small number of bonding electrons become sufficiently agitated to allow them to migrate and become charge carriers. Intrinsic silicon then conducts a little. This naive conceptual model of the conduction behaviour of silicon can profitably be extended a little further. For every negatively charged electron or negative charge carrier that detaches itself from its parent atom and migrates, there is a 'hole' where such an electron should be and this hole is a positive charge carrier since it too can migrate. The hole, or vacancy, can move from atom to atom in the lattice. The behaviour of germanium is much the same as that of silicon except that its electrons become mobile at a lower temperature. At ambient temperatures it has a higher concentration of electrons and holes and therefore conducts more easily, or, conversely, it is a poorer insulator.

In order to fabricate a working transistor a prime requirement is to obtain intrinsically pure semiconductor material. The degree of purity needed is such that impurities are measured in small numbers of parts per several million; refinement to these levels of purity was not possible until about thirty years ago. Such purity is occasionally found in nature: a diamond is a good example and its characteristics depend on its purity.

The refinement technique used for semiconductors stems from the ingenious application of a well-known principle. Moulders have known for hundreds of years that when a casting is made in a mould the solidifying material rejects impurities which are forced into the remaining liquid part. Hence much of the impurities and dirt which get into the molten liquid are concentrated in the risers and pouring funnel when the casting sets. These are cut off when the casting is finished and the metal of the casting is much purer than might have been expected.

In the refinement of semiconductors the material is first refined as far as possible by conventional means. It is then cast into a long cylindrical ingot which is placed in a cylindrical crucible. A radio-frequency heating coil is mounted round the crucible at one end. This heating is localised near the centre of the coil and melts the semi-conductor material at the coil centre. The coil is then moved slowly along the crucible with the effect that the molten section moves with it, the material cooling and recrystallising behind it. The cooling and recrystallising rejects impurities into the molten part with the result that the impurities are swept along the length of the ingot, concentrating at one end. This highly impure end is cut off and returned for conventional refinement. Repeated passes of the heating coil result in practically perfect purification of the semiconductor.

The pure semiconductor material can now be 'doped' with boron or phosphorus. Boron is a tri-valent material, having only three valency electrons; phosphorus is penta-valent, having five. Adding a very small amount of boron to molten silicon causes it first to diffuse evenly through the liquid and then by cooling at the right speed the boron atoms become entrapped in the crystal lattice. Since each boron atom only has three bonding electrons, its neighbouring atoms are not completely satisfied and there is a bond left over causing an artificial hole. Conversely, a phosphorus atom has an electron to spare and creates an artificial negative charge carrier. These artificially produced carriers neutralise some of their equal and oppositely charged counterparts which are generated by thermal agitation thus leaving a preponderance in the crystal of one polarity or the other. The doped semiconductor material so formed is referred to as 'p-type' or 'n-type' depending on the preponderance of positive or negative charge carriers that remain.

A very important aspect of doping is that if an n-type material is further doped with a p-type impurity it first neutralises the n-type impurity already present and then the material develops p-type properties as the concentration is increased. The evenly balanced or neutral material is, of course, not intrinsically pure.

THE MOS TRANSISTOR

Figure C.1 shows an idealised (and enormously magnified) section through a simple MOS transistor. It works quite simply as an electronic switch. If the gate electrode, in the centre, is held at the same potential or voltage as the source, then there is no conduction between the source and drain. When the gate is given a negative bias it causes positive charge carriers to be induced in the n-type material under the gate electrode. Conduction can then take place from source to drain because of the presence of these carriers. The layer of silicon dioxide, although

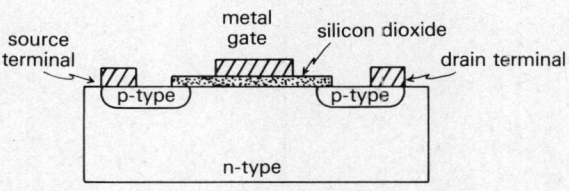

Figure C.1

very thin, is a perfect insulator so there is no current path between the gate and the other electrodes. This is the prime difference between it and the junction transistor and also gives it the properties that make programmable read-only memories possible.

To fabricate transistors of this kind a wafer of n-type silicon has one of its surfaces oxidised to develop a thin film of silicon dioxide all over. The oxidised surface is coated with a material known as photoresist which is a special kind of photosensitive emulsion. It is special in that, when it is exposed to light, its chemical constitution changes and it becomes soluble by an etching fluid whereas unexposed it is not soluble. The resist-coated wafer is clamped to a glass mask, similar to a photographic negative, and exposed to light. After exposure, the wafer and developed resist coating are submitted to an etching process which dissolves the coating where it has been exposed to light through the mask and this leaves windows of silicon oxide unprotected by the resist. Further etching dissolves away the unprotected silicon oxide exposing the semiconductor wafer through windows in the oxide layer.

The wafer is then mounted in a tubular furnace and heated to a precise temperature in the region of 650 °C. Gas which is rich in p-type impurity material is passed over the wafer; the gas reacts with the silicon where it is not covered by the oxide coating and the p-type impurity diffuses into the n-type silicon, first neutralising it, and then forming islands of p-type impure silicon, which become the source and drain. The depth of penetration and the concentration of the impurity depend on the temperature, the gas concentration and the duration of the treatment, all of which must be exactly regulated. When it is appreciated that the furnace is very hot, the gas very corrosive and toxic and the wafer very fragile it is easy to see why it takes great skill and high technology to fabricate transistors at all, let alone get a high yield.

After the p-type deposition the surface is metallised with a thin film of aluminium. This is masked and etched so as to leave only the electrodes.

If the transistor is to be marketed as a discrete device, the wafer is scribed and cut into dice, each containing one transistor. Gold wires are spot welded to the electrodes to connect them to the terminals of the encapsulation. If the transistors are to be parts of an integrated circuit, the metallisation and masking processes are extended to interconnect the individual transistors into the integrated circuit networks and these are then tested and encapsulated.

It is easy to understand the preference for the MOS type of technology in LSI when we appreciate that the fabrication of bipolar devices takes twelve or thirteen successive washing, etching and deposition phases, rather than the four or five required by MOS technology.

The mask making is also high technology: the masks for each stage are first designed and drawn to a large scale. They are then photomicrographed, that is photographically reduced to the minute scale required in the fabrication process. Masks for arrays of devices are prepared by photographically repeating them as arrays on the mask. Storage devices consist principally of arrays of elements and are therefore easier to design and fabricate than are microprocessors which necessarily comprise numerous different networks. It is easy to see, too, why the microprocessor designer chooses a single bus and microprogram store configuration in preference to one with more logical complexity and multiple criss-crossing highways.

The process of fabrication described is merely illustrative. There are now many variants of it and refinements to it. The high technology gets almost unbelievably higher and higher. It involves teams of very clever people from widely varying fields—metallurgy, engineering, photography, physics, chemistry, art—all pooling their skill and ingenuity in the closest collaboration.

Index

Accumulator 12, 94
Addition 12, 19
Address 4, 11, 16, 28
Address modification 34
Addressing 28 ff.
ALU *see* Arithmetic and Logic Unit
ALU Logic 96 ff.
AND element 89
AND function 19
Arithmetic and Logic Unit (ALU) 3, 19
Assembler (program) 21, 56
Assembly directives 58

Backing stores 6
Basic 62
BCD 21, 83
 packed 22
Binary arithmetic 21, 82 ff.
Binary notation 6, 81
Bistable element 93 ff.
Bit 7
Bit-slice configuration 78
Boole, George 86
Boolean algebra 86
Bus 8, *see also* Highway
Busbar logic 99
Busy status 70
Branch 13, *see also* Jump
Byte 4, 7

Central processing unit (CPU) 3, 10 ff., 38 ff.
Central registers 3, 42
Clock pulses 8
 real time 71
Combinational Logic 86 ff.
Compiler program 61
Complement Modulo 2 21, 86
Conditional branch or jump 14, 16, 43
Console 63

Control Logic unit 12
Control Register (CR) 11, 43
Control Unit 4
Converters (A–D and D–A) 66–71
Core Stores 23
Crisis time 73
Cross-assembler 58
Cross-compiler 64
Current page addressing 31

Debugging 13, 59
Decimal adjust function 22, 98
Direct Addressing 30
Division 22, 84

Editor program 65
Emulators 63
EPROM 26
Exclusive OR 90, 97
Extended Addressing 33

Fetch Instruction phase 11
Flag or Machine Status Register 16, 21, 43
Floppy disc 6
Functions (ALU) 19

Gate 7

Handshake principle 70
Hexadecimal notation 82
High Level Language 60
Highway 8, 14, *see also* Bus

Immediate Addressing 17, 29
Index Registers 40
Indexed Addressing 34
Indirect Addressing 32
Input/Output 66 ff.
Instruction 2, 4, 10 ff.
Intel 432 79
Intel 8008 44

Index

Intel 8080 47
Interface units 68
    serial (SIU) 71
    parallel (PIU) 68
Interfacing 66 ff.
Interpreter program 62
Interrupt 46, 47, 49, 70
Intersil IM6100 77

Jump instructions 13, 28

Karnaugh map 91

Latch elements 93, 94
Logic elements 7, 88 ff.
Lovelace, Countess Ada 2

Machine Code 54
Machine instruction 12
Macro-routine 60
Memory reference instructions 5
Microinstruction 8, 14
Microprogram 8, 14
Microprogramming 17
Microroutine 8, 14
Modification 34
Modulo 2 operation 21
Motorola 6800 48
Multiplication 21, 84

NAND element 89
National SC/MP 'Scamp' 50
Negate function 20
Negative numbers 21, 85, 86
NOR element 90

Octal notation 82
Operand 4
Operating system 58
OR element 89

Page (memory) 16
Parallel Interface Unit (PIU) 68

Peripheral devices 5, 66 ff.
Pin-out 8
Pointer registers 40, 41
Program 5
Program Counter (PC) 11, 28, 40
Programmable Read-Only Memory (PROM) 26

Random Access Memory (RAM) 5, 24 ff.
Read-Only Memory (ROM) 5, 26
Register 7
Relative Addressing 31

Shift function 20
Shift register element 96
Signed numbers 21, 85, 86
Simulation program 64
Single-Address Instruction 12
Single-chip microcomputers 74
Stack 35, 36
Stack Pointer (SP) 36, 40
Start–Stop principle 73
Subroutine 59

Texas TMS 9900 78
Tri-state elements 100
Truth table 90

UAR/T 71, 72

Volatility (memory) 24
Von Neumann computer 34

Wilkes, M.V., microprogrammable computer 14
Word (of information) 7
Word length 13, 16, 21, 29
Working store 6

Yield 19, 102

Zero Addressing 35, 36
Zilog Z80 50